BUG
BUSTERS

THIRD EDITION

BERNICE LIFTON

SQUAREONE
PUBLISHERS

The procedures in this book are based upon the personal experience, interviews, and research of the author. Should the reader have any questions regarding the appropriateness of any procedure or material mentioned, the publisher and author strongly suggest consulting a professional pest control operator or public health service.

Because any material or procedure can be misused, the author and publisher are not responsible for any adverse effects or consequences resulting from the use of any of the preparations, materials, or procedures suggested in this book. However, the publisher believes that this information should be available to the public.

Cover Designers: Phaedra Mastrocola and Jacqueline Michelus
In-House Editors: Joanne Abrams and Amy C. Tecklenburg
Typesetter: Gary A. Rosenberg

Square One Publishers
115 Herricks Road • Garden City Park, NY 11040
(516) 535-2010 • (877) 900-BOOK • www.squareonepublishers.com

Library of Congress Cataloging-in-Publication Data

Lifton, Bernice.
    Bug busters : poison-free pest controls for your house
and garden / Bernice Lifton.
        p. cm.
    Originally published: Garden City Park, N.Y. : Avery Pub. Group, © 1991.
    Includes bibliographical references and index.
    ISBN 978-0-7570-0095-9
    1. Household pests—Control. 2. Garden pests—Control. I. Title.
TX325.L543   2005
648'.7—dc22
                                                                2004027291

Printed in the United States of America

10   9   8   7   6   5   4   3   2   1

Contents

PART FOUR

Pests of Property

PART FIVE

Controlling the Controllers

For David, of course.
And to the memory of Rachel Carson,
who first sounded the alarm.

Acknowledgments

Many people have given of their time and expertise to the writing of this book. From distinguished scientists, busy health professionals, conscientious pest control operators, and dedicated employees at all levels of government, I have enjoyed unstinting cooperation. Understanding the need for a book like this, without exception they have offered guidance and support. I'm grateful to all of them.

In addition to writing the foreword to the second edition, the late Professor Walter Ebeling of the University of California, Los Angeles, honored as "the grandfather of urban entomology," read that manuscript for technical accuracy, offering many pertinent suggestions and moral support. His comprehensive book, *Urban Entomology,* still a basic work in pest control, has been invaluable to me in all three editions of this work. His associate, Don Reierson, Staff Research Associate, Department of Entomology, University of California, Riverside, has done yeoman service in reading the manuscript of this edition. I'm deeply grateful to him for his diligence and expertise.

Others who took time to talk with me were Dr.Bernard Portnoy and Dr. Susan Tully, pediatricians at the Los Angeles County-University of Southern California Medical Center, both on the front line of health problems caused by pesticide poisonings. Jerome Blondell of the Office of Pesticide Programs of the Environmental Protection Agency was ever ready with accurate (and depressing) national statistics of accidental pesticide poisonings, the basic rationale for this book.

A special bow to publisher Rudy Shur whose enthusiasm for the project has never failed while shepherding the manuscript through the production process. Thanks, too, to editors Joanne Abrams and Amy Tecklenburg who have helped make *Bug Busters* a more useful book.

The late Bart Andrews, my agent, on a chance phone call, said of my ideas on pest control "That is a mainstream idea," giving me the courage to plunge into the universe that is entomology. I'm forever in his debt.

Lastly, loving thanks to my husband. Without his encouragement and support, I would never have been able to undertake and complete this project for the third time. The book is as much part of him as it is of me.

Bernice Lifton
Pasadena, California

Preface

In the twenty years since *Bug Busters* was first published, there have been some causes for hope that we are getting our pesticide problems under control. That hope is tarnished, however, by stubborn accident statistics that have changed little or worsened since 1985. In 1981, more than 21,000 people, over half of them children, were treated in hospital emergency rooms for pesticide poisoning. In 2000, that number was 23,000—hardly cause for celebration.

Aggravating the problem are ominous episodes in which our passion for trying to wipe out insect nuisances has led us into potentially deadly waters. Here are just two:

1. In April 2001, residues of lindane—a powerful nerve poison for which there is no safe exposure limit—were found in chocolate Easter eggs distributed by major supermarkets in the United Kingdom. The use of this dangerous pesticide is strictly limited in all countries of the European Union. So how did it get into the beautiful Easter candy the kids love? The cocoa came from farms in Ghana, where there is little if any control over pesticide use.

2. In the 1980s and 1990s, methyl parathion, which is legal for use only on certain specific crops and is so powerful that farmworkers are not permitted to go back into treated fields for two days after the pesticide has been applied, was widely used by rogue exterminators in the Midwest and South to control insects in homes. Said one satisfied customer, "I can smell it, so I know it's working." It certainly was. Complicating the danger were mistaken diagnoses by primary care physicians unfamiliar with poisoning symptoms. The doctor for one family with seven sick children thought he was dealing with viral gastroenteritis—in other words, "stomach flu"—and prescribed an antibiotic. Two of the children died. In Ohio alone, 500 homes were treated illegally. More than 1,500 individuals had to be relocated from their homes, at a total cost to the government of $90 million. One pest control operator who was brought to trial for the illegal use of methyl parathion attempted to defend

himself by claiming that he was illiterate and so couldn't read the label on the poison container. That defense didn't wash when a government witness said that the man regularly read aloud from the King James Bible in church.

Now some good news. A new program, *National Strategies for Health Care Providers: Pesticides Initiative,* aims to incorporate environmental issues into the curricula in medical schools. The initiative is sponsored jointly by the U.S. Environmental Protection Agency (EPA), the U.S. Department of Health and Human Services, the U.S. Department of Agriculture, and the U.S. Department of Labor. There is still little education for our doctors on this major health problem, but we can hope for improvement.

And America's children are finding good friends in Congress. On January 4, 2000, the General Accounting Office released a report, done at the request of Senator Joseph Lieberman (Democrat of Connecticut), documenting 2,300 incidences of pesticide exposure in schools. Over 300 of these required medical attention. Said the senator, "We have a national framework for protecting workers on the job, but no such system for protecting children in the classroom. You don't have to be an A student to know that is a double standard that deserves our attention." Two of Senator Lieberman's colleagues, Senator Patty Murray (Democrat of Washington) and former Senator Robert Toricelli (Democrat of New Jersey), shared his views and introduced a bill designated the School Environmental Protection Act (S. 1716). Among other things, this law would require schools to tell parents both when pesticides are being used at their children's schools and which specific chemicals are to be used. Unfortunately, although it was backed by a broad coalition of environmental, public interest, and educational groups, as well as the National Pest Management Association; and was attached as a rider several times to agricultural appropriations bills, this effort failed. But we can hope the senators will try again.

One of the EPA's highest priorities is protecting the health of America's children, and to that end it is encouraging schools to adopt the least toxic strategies to cope with insect and rodent nuisances. The agency is helping school personnel to understand and implement such methods through its publications, by awarding grants to start the programs, and by offering workshops and courses to provide guidance and assistance through partnerships with universities and national environmental institutions. The EPA's Pesticide Environmental Stewardship Program (PESP) joins with state, municipal, and business groups to promote safe pesticide use in our nation's schools, as well as elsewhere. In one outstanding success, New York City's Board of Education reduced pesticide fogging by 90 percent. They had hoped to have eliminated it completely by 2000. Although many localities and school districts now recommend or even mandate that least toxic controls be observed in schools, not all of them follow these practices. Meanwhile, many children still may be exposed to potentially harmful pesticides applied by untrained custodians in their schools. They certainly don't need to face the same risks at home.

Since the first publication of *Bug Busters* in 1985, I have received letters and phone calls from people thanking me for writing this book. This is my major reason for undertaking the project once more. With encouraging new developments in safe and effective pest controls, maybe we will see the grim numbers at the start of this piece start to fall.

How to Use This Book

Bug Busters has been organized in such a way as to help you understand pest control and the problems you will encounter in coping with specific pests. The book is divided into five parts. Part One describes overall strategies that, if consistently followed, will make your home less attractive to insects and rodents. It also discusses the possibility that your real problem may not be pests at all. Parts Two through Five deal with the most common household pests, detailing controls for those most apt to invade your home.

Part Two, Pests of Food, discusses cockroaches, pantry pests, rats, mice, and flies. Part Three explains how to control Pests of the Body—mosquitoes, fleas, ticks, lice, and bedbugs. Not only are all of the pests mentioned in Parts Two and Three unpleasant, but they also can carry serious diseases. (By the way, all can be controlled without chemicals.) Part Four describes Pests of Property—clothes moths, carpet beetles, silverfish, crickets, termites, and carpenter ants. The section concludes with a look at garden pests. Part Five, Controlling the Controllers, details ways of suppressing ants, spiders, and wasps. All of these creatures actually help to control true pests—even though most of us don't want them to come too close. Included in this last section is a chapter on the safe use of chemical pesticides. I firmly believe that sound home construction and maintenance, coupled with reasonably careful housekeeping, makes the use of household pesticides almost completely unnecessary. However, I also know that we are all human and likely to let household chores slide at times and the structure of our homes deteriorate. In addition, seasonal surges in pest populations can bring what seems like an army of insects into your home. If a pest is a known carrier of disease or is especially destructive of property, you will want to bring it under control quickly. Under such circumstances, you may feel an urgent need to use a chemical pesticide. Most pesticide poisonings in the home are caused by the misuse of these poisons or careless storage or disposal of them. Therefore, it is essential to pay close attention and take care if you must resort to using them. Careful reading of this chapter could save you from grief.

There are two toxic substances that I feel can be used safely in the home: Boric acid, an "old reliable," and hydramethylnon, a relative newcomer. Both are among the least toxic antipest chemicals we have, and both are effective against cockroaches. Because those ubiquitous pests are so hardy and persistent, the measures I recommend for most pest infestations—shutting them out, starving them out, and trapping them—may not be enough to get rid of them completely. Both boric acid and hydramethylnon can be placed in out-of-the-way locations where they are no threat to children or pets.

In the chapter on mosquitoes, I also recommend the repellent N,N-diethyl-meta-toluamide (DEET). If misused, DEET is dangerous. However, the risk of a deadly or disabling mosquito-borne disease such as malaria, dengue fever, or encephalitis is much greater than any risk from the correct use of DEET.

Discussions of poisons fall into three categories. Some poisons—for instance, warfarin, a rodenticide—are mentioned simply to let you know that they are commonly used against a particular pest. Others—such as phosphorus, which is sometimes used illegally against rats—are mentioned as being especially dangerous, and I warn you never to try them. The third group includes substances that, like boric acid, are the least toxic control available for a stubborn infestation. Throughout Chapters 1 through 13, the discussions of toxins in this last category have been screened in gray to alert you to the fact that, although the use of a given substance is sometimes warranted, that substance is nevertheless a poison and should be used with caution. It should be noted that toxins in this category have been screened only when the substance is suitable for the average reader to use. Those substances and procedures that should be used only by a professional pest control operator have not been marked with a screen; in such cases, it is the pest control operator who will assume responsibility for proper use of the pesticide. (Information on choosing a qualified pest control operator is provided in Chapter 11.) In addition, in the interest of practicality, those substances mentioned in Chapter 14, Pesticides—Only as a Last Resort, also are not printed with the gray screen, since this entire chapter deals with substances that fall into this third category.

Individual chapters follow a general pattern. The introductory portions about the pest or pests discussed in each chapter (cockroaches, ants, or whatever) describe the appearance, nature, and habits of the animals. The portions of the chapter devoted to ways to control the pests are generally divided into a number of sections, among them the following:

• **Shut Them Out**, which details measures for preventing the pest's intrusion. In nearly all cases, keeping the pest from getting in is the safest control, your first line of defense.

• **Starve Them Out**, which tells you which types of foods and other substances attract the animal. By making your home an inhospitable environment, you can eliminate a great many types of pests.

• **Wipe Them Out**, which tells you how to eliminate the survivors.

Not every chapter has information in each category—for example, in Chapter 12, Plant Pests, most of the above measures are not applicable, so the emphasis is on control.

I hope that this organizational structure will make *Bug Busters* truly helpful, a source book of environmentally sound pest control methods that you can refer to repeatedly over the years and that will make your home a safer place.

PART ONE

No Pests, No Poisons

1. Controls, Not Chemicals

H ave you ever set off a bug bomb in your home and then felt sick for a couple of days? Or spread snail bait among your seedlings and prayed that the dog—or the kids—wouldn't sample it? Maybe you've heard about the woman who was about to go away for a weekend and leave her beloved cat home. It seemed like a good time to get rid of the cockroaches in her apartment. She set out plenty of water and food for the cat, sprayed the kitchen thoroughly with a common household pesticide, and locked the door behind her. When she came home two days later, all the cockroaches were dead. So was the cat.

If so, you've probably wondered, "Aren't there safer ways to fight pests?"

THE DRAWBACKS AND LIMITATIONS OF CHEMICAL PESTICIDES

Modern pesticides usually bring quick death to those unpleasant little creatures that are determined to move in with us, but what do you do when the cockroaches—or fleas or snails—show up again and again, as they are apt to do? Keep laying on the poisons?

A slow, searching walk along the pesticide aisles of a hardware store or garden shop can be a disturbing experience. From the acrid odors, whether produced by the insecticides or the agents used to disperse them, you know instinctively that these products can do you no good. Your instincts are right. As one chemist once said, "If you can smell it, it will probably do some harm."

Scanning the package labels can make you even more uneasy:

• Do not breathe spray mist. In case of eye or skin contact, flush with plenty of clear water. Contains a cholinesterase inhibitor (this is a substance that impairs nerve function, but the label doesn't tell you that).

• Bait may be attractive to dogs. Toxic to birds and other wildlife. Keep out of lakes, streams, and ponds.

And always,

"Keep out of reach of children."

If these materials interact with air, you could be breathing toxic vapors as long as residues are left, which could be indefinitely.

What's worse, chemicals aimed at particular pests may have been found to be unsafe by the U.S. Environmental Protection Agency (EPA), yet are allowed to remain on the market shelves until stocks are gone. Confused consumers rely on retail sales help for information. Unfortunately, these people generally know as little about these chemicals as the consumers themselves do.

Actually, everyone is vulnerable to these toxins—the home gardener who accidentally sprays a bug killer into the wind; people with allergies, respiratory ailments, or skin problems; older adults; young children; and pets.

Okay, so you're young. You have neither children nor pets. Your skin and lungs are in great shape. You have no allergies. You are also absolutely sure that you *always* handle hazardous substances without ever endangering yourself or anyone else. The following fact, however, may give you pause: Some of our most destructive pests are now resistant to nearly all the chemical killers we have. Hundreds more are immune to at least some pesticides.

Among the species that readily bounce back from pest control chemicals are those that are most dangerous to us—flies, fleas, mosquitoes, ticks, and lice. Other resistant insects include major pests of agriculture, forests, and stored foods. Since chemical pesticides were first used, in the mid-nineteenth century, there have been more than 500 reported cases of insects developing resistance to the insecticides meant to control them. Today, more than 500 species have developed a resistance to at least one insecticide, and many pest species are now resistant to almost all of the poisons in our arsenal. We may seem to win a few battles here and there with insect pests, but we are beginning to realize that we have lost the war. Although the use of insecticides on crops in the United States has increased tenfold in both amount and toxicity since 1945, the share of crop yield lost to insects has nearly doubled in the same period. Despite the regular drenching of our soils with powerful poisons, insects still devour about 20 percent of our farm products.This is roughly the percentage American farmers suffered in 1900, fifty years before the first onslaught of organic pesticides began. And we all pay the price in higher grocery bills and possible residues on food. Either way, the picture is hardly encouraging.

Since commercial agriculture now relies so heavily on toxic chemicals to make and keep its products marketable, we certainly do not want to add household pesticides to the brew of poisons we already take into our bodies. How deadly are these substances? The world found out on December 2, 1984, when a toxic gas, methyl isocyanate (MIC), a basic ingredient in the common pesticide carbaryl (sold under the brand name Sevin), leaked from a Union Carbide plant in Bhopal, India. As many as 2,500 people were killed outright, and another 60,000 were seriously injured, many of them disabled for life.

So chemicals may not solve your problem at all, even in the short run. More worrisome, homeowners who pour unused pesticides down a drain or drench garden soils with them are major polluters of our water. They are also contaminating our seafood. Shellfish taken from the coastal waters off southern California have been found to contain dangerous levels of pesticides. Urban streams on Oahu are so badly polluted with the insecticide dichlorodiphenyltrichloroethane (better known as DDT)—a long-lasting chemical that persists despite having been banned since 1972—and chlordane—banned in 1988—that Hawaii's Department of Health is still telling people to avoid eating crayfish, tilapia , crabs, even o'opu taken from local waters.

If you fall into the trap of using more and more pesticides, you run the risk of making yourself sick or developing an allergy to them. Meanwhile, the pests you are trying to kill, hardly fazed by the same substances endangering you, may very well keep coming back. As our forests shrink and wood becomes more expensive, pressed wood (which contains formaldehyde) is increasingly being used to make furniture. Formaldehyde is a known sensitizer that makes people more vulnerable to danger from pesticides. If we add to these hazards the pollutants from automobiles and industry that are already infiltrating our bodies, doesn't it make good sense to keep our personal space as free of contaminants as possible?

After biologists like the late Rachel Carson discovered that chlordane and DDT, both organochlorines, were deadly over time to both humans and animals, we began to use a group called organophosphates, which were considered to be safer than the organochlorines. Both groups, however, contain extremely dangerous toxins and some that are relatively mild. *None should be used carelessly.* (For a more complete description of these pesticide groups, see page 250 in Chapter 14.) Yet people are still being hurt by these poisons. There is now a recognized medical condition called *multiple chemical sensitivity (MCS).* The so-called safe alternatives have now been documented to cause a multitude of problems, including very serious central nervous system problems. Possible effects of prolonged exposure include an impaired ability to concentrate and to solve problems, loss of memory, slurred speech, slowed reaction times, flaccid muscles, and poor dexterity. Because people who suffer from this syndrome lose their capacity for meaningful work, many of them become depressed and demoralized.

MCS is often misdiagnosed by doctors who are unfamiliar with the condition. As a result, people who suffer from it are sometimes thought to have emotional problems and are shunted off to psychiatrists, who really can't help them. One victim of this syndrome, a former student at Harvard Law School, became violently ill after exposure to a common household insecticide. She suffered damage to her heart as well as to her nervous, reproductive, and immune systems, and is now allergic to many chemicals. At last report, she was living in a tent on unsprayed land in Texas. Her leave from Harvard has run out, and she has little hope of resuming any career.

OTHER WAYS TO FIGHT PESTS

Are there other ways to discourage and drive off the insects and rodents that find our homes and gardens attractive—ways that work permanently? Yes, there are. Our ancestors knew how to get rid of all sorts of unwanted wildlife, both indoors and outdoors, without poisons. Two of their most powerful weapons are yours for the taking—cleanliness and sunlight. Other methods they used were often not much more complicated than these. And best of all, pests cannot become immune to these safe, effective techniques.

Insects and rodents have been our unwelcome companions ever since humans first moved into caves, put on clothes, and learned to grow and store food. In making life more secure and comfortable for ourselves, we also made it more secure and comfortable for many species of little wildlife.

Few of us want to move back out under the trees and run naked to get rid of the pests, and most of us would find a diet of roots and berries pretty dull. Down through the years, however, observant individuals have noticed that insects and rodents avoid certain substances or conditions that humans find harmless or pleasant. They also observed that if the little creatures lacked food or water, they moved out.

Our ancestors, wherever they lived, stored food against the inevitable times of shortage. Some stored their stockpiles in holes lined with straw to keep insect raiders out. Some Chinese people still build small shelters of clay and soft mortar to safeguard their grain, as their forebears did thousands of years ago. Early peoples also mixed their grains with aromatics like bay or eucalyptus leaves, or inedibles like sand, ashes, or sulfur, to repel pests. Farmers in Galicia, Spain, still build small granite and wood huts, called *horreos*. Unchanged in design since about 500 B.C., these granaries are almost completely rodent-proof. In India, grains were once spread on rugs placed in direct sunlight to drive out light-fleeing insects.

Many of the pest control methods described in this book are updated versions of these time-tested techniques. They may have kept your grandmother's house pest-free and given it a pleasant fragrance that you remember to this day. In today's enthusiasm for instant solutions, we tend to ignore safer, if somewhat slower, ways of controlling pests. Yet these less drastic methods still make good sense. After all, you wouldn't go after a gang of neighborhood burglars with a bomb if strong locks, good lighting, shrill alarms, and an alert police force could do the job effectively year after year.

That said, not all of the techniques I will describe here are equally effective in all situations—but then, neither are the chemicals. Climate, seasonal weather, natural surges in pest populations, soil conditions, and your own thoroughness and persistence may all affect how well these measures work for you.

UNDERSTANDING AND CONTROLLING PESTS: A NEW PERSPECTIVE

When that comfortable home you now enjoy was built, it displaced thousands, possibly millions, of tiny animals, all of which were well integrated into an environment with adequate food and water, as well as safe nesting sites. To the extent that your home provides these basics of life, insects and rodents will try to occupy it. Indeed, your home may actually be a better environment for them than the outdoors. Besides providing a huge supply of food and water (by insect standards), most houses have innumerable nooks and crannies in which small creatures can hide from predators, including humans.

Like all displaced animals, your land's former occupants need a new home. Unless you are unusually hospitable, you probably don't want to share yours with the many-legged. So how can you get them out—and keep them out—of what is now *your* turf?

The following chapters will describe the least toxic measures available for controlling the most common pests that want to eat your food, clothing, furniture, and plants—even you. Many of these methods have been known for generations, and the new technology of integrated pest management (IPM) has developed, and continues to develop, sound new ones.

Originally designed to help farmers, IPM is a strategy that uses technical information, ongoing monitoring of pest populations, crop assessment, and other techniques to control pests. Chemicals are just one part of the strategy, and their use is kept to a minimum. In the 1980s, many pest control professionals were bitterly opposed to the concept of IPM, fearing that it would deprive them of their arsenal of chemical weapons. But since then, there has been a sea change in the pest control industry's attitude. Today, IPM is thought by many pest control professionals to be the only way that we humans can win our struggle against destructive insects and rodents without damaging ourselves, our domestic animals, or our environment. In fact, the trade association for pest control professionals has even changed its name from the National Pest Control Association to the National Pest Management Association.

Insects—Awesome Fertility, Phenomenal Resistance

First of all, you need to accept the fact that you are engaged in a war that you simply cannot win. There are just too many insects. Estimates of the number of *kinds* of insects on Earth vary from 750 thousand to 30 million distinct species. *Individual* insects are beyond counting. Scientists can make only picturesque guesses, such as the following:

• The total weight of all the insects on Earth is far greater than that of all other species combined; or

• The progeny of two flies mating in April would, if all survived, cover the earth by August with a disgusting blanket of flies forty-seven feet thick.

Other insects have a reproductive ability just as awesome. The females of some species lay a million eggs each in their lifetimes—eggs that, in many cases, can hatch months or even years later. What is even more unsettling is that many pests can become immune in just a few generations to the most deadly insect poisons chemists have devised, and insect generations are often measured in weeks. (Rodents, which produce only several litters a year, also are showing resistance to once-lethal substances.)

Today, resistant houseflies are hardly slowed by DDT, and one common pantry beetle has been found thriving on a diet of belladonna, aconite, and strychnine, all substances that are fatal to other living creatures. Not surprisingly, no insect species known to us has ever been completely exterminated. They are the most hardy and adaptable creatures on Earth.

In the 1970s, when chemical pesticides were almost always the weapon of choice in dealing with unwanted insects, the number of resistant species nearly doubled, growing from 224 to 428. In that same decade, the number of cases of malaria, a mosquito-borne disease that kills between 2 and 4 million people (most of them children) every year, also doubled, as its carrier, the *Anopheles* mosquito, became increasingly immune to our chemical arsenal. More than 500 species of insects and mites, 270 weed species, more than 150 plant pathogens, and about a half-dozen species of rats are now resistant to the chemicals that once controlled them. Even more discouraging, immunity to more than one pesticide and to more than one class of pesticide is increasing by leaps and bounds. There are more than 1,000 insecticide combinations to which some pests are immune, and at least seventeen insect species that shrug off all classes of insecticides.

Clearly, our insect and rodent enemies are invincible. However, with vigilance and a sound strategy, you can reach an acceptable stalemate. By using the controls described in this book, you will also lower your risk of accidental poisoning.

Basic Antipest Strategy

Your basic strategy for getting rid of small freeloaders is simple. You need to find out:

- How they are getting in;

- Where they get food and water; and

- What conditions kill them or drive them off.

Once you determine these things, launch your antipest campaign by:

- Shutting them out;

- Starving them out;

- Rotating your food, clothing, and furniture;

- Zapping them with very high or very low temperatures;

- Exposing them to sunlight and fresh air;

- Using fragrances they dislike; and

- Trapping them.

In the rest of this chapter, we will look at some of the basic principles behind each of these approaches.

SHUT THEM OUT

A tightly built, well-maintained house is essential in any pest control campaign. An uncapped, unused chimney gives easy entry to rats, squirrels, flies, spiders, and mite-infested birds. Defective roofing can have the same effect, while gutters and downspouts choked with dead leaves are ideal insect nurseries. Broken windowsills, leaky plumbing, and loose flashing around chimney and vents make for damp walls and the mold that many insects feed on.

Small cracks in walls, both inside and out, give cockroaches, ants, rodents, and other pests easy access to the interior of your home, as well as safe hiding places once they are inside. Most pests need only a tiny fraction of an inch to establish themselves in a house. Cracks or small openings leading to attics or wall voids allow bees or wasps to establish colonies in inaccessible places in your home. Damaged air grilles in attics and basements are another likely route of invasion. Openings around plumbing pipes are the most frequent entry for cockroaches. (See Figure 1.1.)

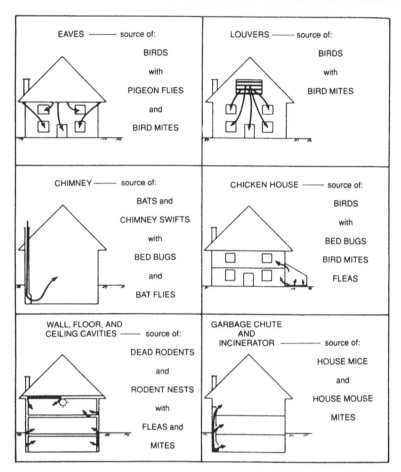

Figure 1.1 Common Sources of Bird-, Bat-, and Rodent-Borne Pests

Source: Household and Stored-Food Insects of Public Health Importance and Their Control, Home Study Course 3013-G, Manual 9 (Atlanta: U.S. Centers for Disease Control, 1982.)

Inspect and Repair Your Home

Your first step in shutting out pests is a careful home inspection. You may need professional help here—a qualified roofer and a sheet-metal contractor to make accurate assessments of the state of your roof, gutters, and flashings; a carpenter to point out needed repairs of windowsills and door frames. You can probably take care of small cracks in the outside walls on your own with a caulking gun, latex paint, or patching stucco. Tight window screens in good repair and screen doors that close automatically and quickly are your best defenses against flying pests.

As any slum-dweller can tell you, broken toilets and rusted-out floor drains can provide a way for rats to get into a run-down building. So make sure that all of your toilets are tightly bolted to the floor and that they work properly. The strainers covering drains in laundry tubs, showers, bathtubs, and basement or garage floors should be strong enough and of fine enough gauge so that no rat or cockroach (both of which are sewer-dwellers) can push its way through them.

Modify Your Landscaping

Look at your landscaping with a skeptical eye. Though they can make a home visually attractive, shrubs and flowers hugging the outside walls can be a staging area for an invasion of mites, pantry pests, or spiders. A clear strip that is at least eighteen inches wide, made of either concrete or sand and circling the foundation, will act as a barricade against most crawlers and make them easy marks for their natural predators. If you have Algerian ivy or any other broad-leaved ground cover bordering your lawn, pull it up. Broad leaves growing close to the ground are a perfect habitat for black widow spiders, rats, mice, and cockroaches.

Eliminate Dampness

Your home may be as tight as a drum yet still attract insect pests. If this is the case, interior dampness is probably the cause, and it can come from several sources. In a structure under thirty years old, the problem may be built in. With hollow-wall construction, which is now common throughout the United States, moisture that can come from air being blown through the house via either air conditioning or central heating condenses in the wall voids. In time, fungi can grow in such an environment, to be followed by mold-eating insects. From these secluded feeding grounds, the insects can easily find their way into your living quarters through vents, electrical outlets, and wall switches. Today, realtors consider mold almost as serious a problem as termites.

In addition, rooms in today's houses tend to be small and relatively airtight, with less air to absorb and disperse the moisture. In many parts of the country, houses are built on concrete slabs and lack basements or any subfloor ventilation. The slabs themselves, made of cement and water, gives off moisture long after they have been poured. And sliding glass doors admit garden crawlers such as millipedes, centipedes, and sowbugs, little animals that find their way in but can't find their way out. Enclosed patios that are lush with foliage in wooden or brick planters also draw moisture-loving insects. The planters themselves, unless waterproofed on all sides (thus sheathed from the soil below) and set against crack-free walls, are potential gateways for termites. Sprinkler-drenched stucco and damp crawl spaces also favor subterranean termites.

Our passion for cleanliness compounds this problem. Dishwashing, bathing, and laundering for a family of four add about three gallons of water to the immediate environment every day, a total that can easily be tripled by a high water table and lawn-watering.

If you think your home is too damp, you should consider buying a dehumidifier as a major pest control step. In less severe conditions, using an air conditioner or even a regular fan also can dry the air.

STARVE THEM OUT

Now that your home is as pest-tight and dry as possible, you can begin to starve and clean out those raiders who were there before you began your campaign or who managed to slip by your barriers, as some occasionally will. Cleanliness and proper food storage are your strongest weapons here. Regular vacuuming of carpets, upholstered furniture, draperies, and shelves sucks up adults, larvae, and eggs, along with any food crumbs that could sustain them.

If you don't have an automatic garbage disposal and garbage collection in your community is inadequate, your trash barrels are probably drawing and incubating flies. Become a political activist; join with your neighbors to fight City Hall and demand more frequent pickups. And until service improves, wrap food garbage in plastic bags and close the bags tightly before putting them outside. Then deodorize your garbage cans with a weekly rinsing of hot, sudsy water and borax solution.

How about the area around your washer and dryer? Are there lint fluffs near the dryer, damp spots under the tub and washer, and bits of soap here and there? Warm, sheltered dryers are favorite nesting sites for rodents, who use the lint to keep their young cozy. Tiny amounts of water satisfy thirsty roaches, who also find soap a tasty meal. So vacuum up all the lint (be sure to move the dryer away from the wall), clean up any and all scraps of soap, and make it a point to keep the laundry area dry.

Spring Cleaning—Still a Good Idea

Do you remember your mother's or grandmother's annual flurry of spring cleaning? Her main reason for all the dusting, rug-beating, and corner-scouring may have been to freshen the house after a winter of stuffy, heated air. But whether she knew it or not, she was also disrupting the reproductive cycles of house pests who had laid their eggs the summer and fall before. Such interruptions seem to slow the biological clocks of developing insects.

She sunned the rugs for hours on a clothesline and then thrashed them with a bamboo beater. She took down the heavy draperies and brushed out all the folds (safe havens for tiny creatures) before storing them in an airtight chest for the summer. She emptied dresser drawers and dusted them, cleaned kitchen shelves and changed their paper lining. All the bustle left the home fresh and airy, as well as pleasantly tidy. More important, her intense housekeeping activities were excellent pest preventives.

This annual rite of spring seems to have faded away in recent years, but if you do it—whatever the season—it will boot out unwanted guests.

If the freshening-up sets you to thinking about redecorating, consider taking up your wall-to-wall carpeting—a good hangout for fleas, carpet beetles, cockroaches, and their ilk—and replacing it with area rugs. These are easy to lift for thorough cleaning, and the wood or vinyl flooring around them offers little comfort to a tiny intruder looking for a hideout.

Store Food Properly

How do you store your staples? Keeping dry foods in bags and cardboard boxes is like displaying a menu for grain- and legume-eating pests. Tightly closed metal, glass, or plastic containers lock these nuisances out, and separating new stocks of food from the old prevents the spread of any possible infestation. If you store anything long enough, something will come along and eat it, so buy your groceries in amounts that you are likely to use up in a short time.

Clear Out the Clutter

A cluttered home attracts insects and other small creatures. Its undisturbed corners offer them innumerable safe havens so, of course, they'll move in. Soon their relatives and descendants show up. If you're an incurable "saver"—always keeping string, boxes, old newspapers, and bits of this and that against some possible need down the road—try to break the habit. Paper and cardboard, long-forgotten stuffed toys, and clothing that has been unworn for years all eventually draw wildlife. Throw out the junk and give anything that is still usable to a service organization that recycles such items.

Incidentally, while you are cleaning all those neglected places, be sure to wear sturdy work gloves in case you meet up with a spider that is resentful of your intrusion.

Even after your house is no longer attractive to the most obnoxious pests, continue to air out your rooms and expose them to sunshine frequently to make any unwanted guests feel unwelcome.

SHAKE THEM UP

Rotating your clothing so that all garments are worn often and then cleaned or aired is one of your best protections against moths, silverfish, and other fabric-fanciers. Fashion experts advise that if you haven't worn a particular garment in two years, you should give it away. That's good advice—whether you want to be stylish or just want your clothing to be free of moth holes.

When you wear woolens, for example, you disturb any moth larvae that may be lurking in the folds, brush them or shake them off the fabric, or expose them to shriveling sunlight. When you launder the article, if there are any larvae remaining, you drown them. Dry cleaning, which uses chemicals, also kills them. Who needs an additional chemical?

Do you enjoy rearranging the furniture? You may do it to be creative, but any time you do this you are also making life unpleasant for resident insects. Exposing carpets and floors that have been underneath heavy furniture makes it easy to vacuum up carpet beetles and other undesirables. (If you're like me, though, and don't ever rearrange furniture once the movers leave, you may be asking for trouble.)

Rugs and carpets under low-slung heavy chests, sofas, and beds are hard to reach with most vacuum cleaners, so moth larvae, carpet beetles, and other textile-loving pests can gather there. And while you're at it, don't forget to rearrange the pictures on the walls, because bedbugs and cockroaches can set up housekeeping behind a frame and under a picture's paper backing.

Take Control of Your Garden

Controlling garden pests need take no more effort than fighting those in your house. Obviously you can't entirely build pests out of a garden, but some of the techniques that work indoors, such as good sanitation and mechanical barriers, can be effective outside as well.

You can apply other strategies to a garden also. For example, avoid planting large stretches of one type of plant or shrub. Having many plants of the same variety in a single location spreads a banquet for insects that feed on that particular vegetation, and is the main reason many of today's farmers still depend on chemical pesticides. You might consider planting an English-type garden instead, with colorful profusions of different flowers and shrubs. Timing the planting and harvesting of vegetables so as to avoid their insect pests is another way to outwit the freeloaders.

Your best aids outdoors are your wildlife allies—birds, lizards, bats, toads, predatory insects—who feed on insects at all stages of their lives. One of the biggest drawbacks of chemical pesticides is that they kill friend and foe alike; and when the predators—which are much more vulnerable to the poisons than their prey—are gone, pests can return in greater numbers than before since you have killed off nature's most powerful controllers. Making matters worse, insects that were formerly no problem at all, when suddenly free of their natural enemies, can become major headaches. One of the clearest examples of this is spider mites. Once minor troublemakers, these creatures have now become a worldwide threat to forests and agriculture as their predators—ladybugs, predacious mites, and pirate bugs—are being decimated by pesticides. Tragically, these predators take much longer to reproduce than do their prey.

WIPE THEM OUT

Wiping out insect and other types of household pests does not mean having to use poisons. There are much safer approaches that can have the same effect.

Roast Them Out or Freeze Them Out

Long before anyone dreamed up chemical insect-killers, it was known that very high and very low temperatures kill most household pests. Few survive prolonged exposure to subzero conditions or heat of 140°F. By keeping candy and grain-based foods in your freezer or refrigerator, you can protect them from the multi-legged marauders. Before World War II, European pest control operators regularly blew air heated to 152°F into apartments to clean out infestations. You can apply the same principle in your own home. However, although spontaneous combustion occurs only at temperatures of 400°F and above, common sense prohibits leaving an overheated house untended. There could be danger of electrical malfunction if your house's wiring cannot accept the high current needed to boost the heat. A good time to do this is before you move in, so that there is no danger of damage to valuable furnishings.

Trap Them, Barricade Them, Repel Them

If you know what a specific pest likes to eat, you can set up a trap that is harmless to humans and pets, but lethal to the insect you are struggling with. For example, a simple trap baited with meat or syrup can kill hundreds of flies a week; a small jar half-filled with beer finishes off cockroaches by the dozen; and silverfish can be done in with a jar partly filled with flour. Cockroach and silverfish traps should be set along the junctures of walls and floors or walls and shelves, where the insects usually crawl.

If you would rather keep a pest away, try a simple barricade. For example, a smoothly finished removable metal grid placed level with the pavement outside your doorway will stop any crawlers headed inside. A snail or slug can't inch past sawdust. Caterpillars can't negotiate a band of hay girdling a tree. Petroleum jelly stops ants dead in their tracks. By thinking along the same lines, you could come up with solutions as simple and effective as these.

What About Herbal Controls?

You might want to explore using herbs to repel unwanted wildlife. Many insect pests have keen senses of smell and avoid sharp scents like those of mint, tansy, and basil.

Known since ancient times for their insect-repelling odors, herbal controls have been used seriously since the early nineteenth century. Although they will not do much if your housekeeping is sloppy, some plants do help hold off insects if combined with good home maintenance.

Since 1942, the U.S. Department of Agriculture's Center for Medical, Agricultural, and Veterinary Entomology (CMAVE) in Gainesville, Florida, has tested hundreds of herbal derivatives. A few show up well. Among the more effective are oil of spearmint, oil of pennyroyal (a common roadside plant in the eastern United States), and citronella. As repellents applied to the skin, however, none of these works as well as the chemical N,N-diethyl-meta-toluamide (better known as DEET). In addition, oils of pennyroyal and spearmint can cause skin rashes.

Nor can herbal products protect large spaces. They have little long-distance effectiveness, and in such conditions work about as well as a smudge fire.

Some plants, however, are known to repel particular pests. For example, in the southern United States, many people plant wax myrtle around the foundations of their homes to drive off fleas that cats may carry into crawl spaces.

Much work on the use of herbal products as insect repellents is going on in India, where the Bhopal disaster that killed thousands in 1984 is indelibly printed on people's memories. Moreover, chemical pesticides are too expensive for widespread use in that country. Plants like the garlicky-smelling neem tree seem to hold promise for good pest control. Lately, neem extracts have been showing up in American garden shops, and even major chemical manufacturers are showing interest in low-impact herbal and natural repellents.

What is a commonsense approach to herbals? If you hear of a plant that has helped a friend to control a particular pest that is now bugging you, try using that plant along with the methods recommended in this book. You shouldn't expect miraculous or immediate results, but you could achieve an acceptable level of control.

Practice Patience and a Little Tolerance

If you tend to panic and grab automatically for a spray or fogger when a bug comes near, try to get that knee-jerk reaction under control. The chemical could do more harm to you than to the insect you're trying to kill. In fact, that little animal may not even be a pest at all. Only about one-quarter of 1 percent of all insect species actually qualify as pests.

Some scientists believe that without the arthropods (the catogory of animals in which insects, spiders, and mites fall), the human species would die out. Most species are beneficial. They turn decaying matter into plant food and, by helping plants to reproduce, they feed us, keep our earth green and beautiful, and cleanse our air. The threads of that silk shirt you treasure were spun by moth larvae. We get shellac from the secretions of a scale insect that infests the ficus tree. And lately science has been finding the bugs helpful for research purposes. By observing the length of time it takes for the larvae of the flour beetle to develop when feeding on specific varieties of cereals, nutritionists can measure the food value of new cereals. The chemicals that fuel the lightning bug's lantern show great promise for the accurate diagnosis of bacterial infections of the urinary tract. In the never-ending warfare among their kind, a great many insects destroy the pests that would destroy our food or otherwise endanger us.

Can insects even be helpful in tracking down murderers? Actually, yes. Using a new pest control specialty, forensic entomology, trained professionals analyze the kinds of insects that are attracted to human corpses (which to them, after all, are carrion). Their larvae and pupae show up and, by correlating this with relevant weather data, researchers can accurately determine a person's time of death. Such precise detective work can be used to wipe out a homocide suspect's alibi or confirm another person's innocence. And who would want to kill a graceful swallowtail butterfly or soaring dragonfly (which, incidentally, is a major predator of mosquitoes)?

Ultrasonics and Electromagnetics

You have probably noticed advertisements proclaiming that a particular electromagnetic gadget or an ultrasonic gizmo will solve all your pest problems. A simple device that is free of dangerous chemicals and environmentally friendly, and that costs anywhere from $40 to $130 each—just switch it on and the pests will vanish. Don't you believe it!

The marketers writing the ads make these things sound like the most effective control for household insects and other pests, and may even claim that their products have been scientifically tested. What they don't tell you is that the tests were seriously flawed. These contraptions are useless—in fact, worse than useless, since they make householders think that they don't have to do anything else to clean out the wildlife. Over the past twenty years, many valid scientific studies sponsored by the EPA, cooperative extensions at state universities, and the governments of other countries have shown these things to be worthless. As far back as 1980, the EPA stopped the sale of these bogus pest control devices. In two cases, termites had built runways on "repellers" being tested in the field. Recently, however, the deceptive ads have been popping up again, with equally preposterous claims. The marketers' loophole is that they don't claim that the devices will repel "insects," just unnamed "pests." And so the ads still show up. You are throwing your money away if you buy any of these products thinking that they will get rid of the nuisances. The devices may be marked "EPA," followed by a series of letters and numbers, but this indicates only that the manufacturer has told the EPA that they are in business. The EPA is not—repeat *not*—endorsing them.

For a time, ultrasonic devices seemed to make more sense. Rats and mice have intensely acute hearing. Any loud noise sends them bolting for escape. Environmentalists and food processors once thought such devices held great promise for rodent control. Sad to say, that promise has not been fulfilled. Ultrasound is directional, does not penetrate stored objects, and loses intensity very quickly with distance. And within a matter of weeks the rats and mice get used to the gadgets and return to their accustomed feeding places.

The one verifiable fact behind these devices is that a mouse can be killed by ultrasound generated at extremely high frequencies and decibels, both of which are dangerous to human beings. It is very hard to kill a rat this way. In addition, such high frequencies use a great deal of energy. Tests run by the Division of Agricultural Sciences of the University of California have all proven negative.

The following, quoted by the U.S. Centers for Disease Control and Prevention (CDC) in their manual on rodent control, sums up science's evaluation of ultrasound as pest control:

> Ultrasound will not drive rodents from buildings or areas; will not keep them from their usual food supplies and cannot be generated intensely enough to kill rodents in their colonies. Ultrasound has several disadvantages: it is expensive, it is directional and produces "sound shadows" where rodents are not affected and its intensity is rapidly diminished by air and thus of very limited value.

The rodents themselves have passed the ultimate judgment on ultrasonic repellers. California state health official Minoo Madon even found a family of mice nesting in one!

The bottom line is that *sanitation,* not gadgets, is the foundation of pest control. And sanitation means eliminating filth, both food waste and human waste. It also means eliminating clutter, which provides harborage where these small animals can find shelter. If you find cockroaches touring your kitchen sink or hear the patter of rodents' paws in your walls or attic, ask yourself:

- How did they get in?

- What are they finding to eat and drink?

- Where can they hide?

- What are the safest ways to get them out?

The steps you take in answering these questions and making the needed changes will get matters under control.

Ultimately, you may have to apply a chemical—but not nearly as much or as often as you would if you hadn't first made your home less attractive to the invaders. Indeed, with reasonable care, you may never have to use a chemical pesticide again.

2. Is Your Problem Really Insect Pests?

In her dramatic and groundbreaking book *Silent Spring* (first published by Houghton Mifflin in 1962), marine biologist Rachel Carson warned of the dangers of our heavy reliance on chemical pesticides. She told the story of a woman who was deathly afraid of spiders. Finding a few in her cellar one August, she sprayed the whole area thoroughly—under the stairs, in the storage cupboards, around the rafters—with a solution of DDT and petroleum distillates (liquids derived from crude oil). Wanting to make sure she had killed every last weaver, she repeated the spraying twice more at intervals of a few weeks. After each go-round, she felt nauseated and extremely nervous.

Following the third spraying, more alarming symptoms appeared: phlebitis, fever, and painful joints. Shortly after, the woman developed leukemia and died.

Although the real culprit in this case may have been the petroleum distillates and not the DDT, the tragic irony of this woman's suffering and death is that the spiders terrifying her were in fact keeping her cellar free of insect pests. They may even have been relatively immune to the substance she was spraying. Furthermore, she could have cleaned them out more effectively—certainly more safely—with her vacuum cleaner.

This woman was a victim of entomophobia (an irrational fear of insects). While hers was an extreme case, revulsion caused by insects and other tiny animals is widespread. Many of us pull back in fear when an odd-looking beetle crosses our path. We know little if anything about it, yet we are repelled. True, it might eat things we value, such as our garden plants or our food. But it is actually more likely that the creature helps us by serving as food for animals that are useful to us, such as birds and lizards, or by preying on even smaller beings that destroy plants, such as aphids and scales. In addition, it may aerate the soil, making it possible for us to grow healthier plants.

Most of us cannot predict or control an insect's behavior. Unlike most other wild species, insects don't seem to be afraid of us. Indeed, they ignore us unless we are about to crush them with our feet. Is that what seems so threatening—that we have almost no impact on another living being?

ENTOMOPHOBIA

Some authorities claim that as many as one person in one hundred is terrified of insects and spiders, a phobia (a deep-seated fear that is triggered by a specific thing or situation that actually poses no objective danger) that they probably learned from the people around them. Some of this is even embedded in our language. If we see an ill-kept dog, we call it a *fleabag*. Markets replete with cheap merchandise are called *flea* markets. If someone plays an unpleasant trick on us, we call it a *lousy* trick. When we're sick, we say we feel *lousy*. If someone is bothering us, we say "Quit *bugging* me." A balky computer is said to have *bugs*.

Yet although insects can get along very well without us, we cannot ignore them. They were around hundreds of millions of years before the first humans showed up on this planet and, from all indications, they will probably still be here long after we have vanished. But while we are here together, we humans depend on insects for both our food and a livable environment.

Not everyone who is afraid of insects would carry this to the point of slow suicide, but even educated individuals can work pretty hard at it. A retired elementary schoolteacher friend of mine recently announced that she had just installed an electronic bug-zapper in her garden, although she had already been regularly dousing her garden with chemicals. When I told her that she was also killing insects that would hold down the pests, she grimaced in disgust. "I hate bugs," she said, "*all* of them." One wonders how many children she has infected with her fear.

Actually, her bug-zapping gadget attracts more winged insects than would probably come to her garden if she didn't have it. Meanwhile, an evening's conversation on her patio is punctuated by the crackle of raw electricity as a living creature is electrocuted every few seconds.

Psychologists aren't sure what causes phobias. According to the *New Columbia Encyclopedia*, phobias arise from inner conflicts that may be rooted in childhood insecurities. In addition to the fear of insects, common phobias include the fear of heights, enclosed spaces, birds, cats, and even other people. A person with entomophobia who is constantly spraying for insects—even those that are harmless or merely annoying—may be demonstrating a kind of defense mechanism in an effort to overcome anxieties caused by deep-seated emotional distress.

Overcoming Fear of Arthropods

A person who has an intense fear of insects or other small animals *can* overcome these rampant feelings. Suppose, for example, that you are terrified of spiders. The following steps can gradually help you to build up your tolerance of them:

1. Spend some time looking at pictures of spiders.

2. Read about spiders.

3. Get up early in the morning and watch a wheel net spider spin a web on one of your garden shrubs. You will see her build one of nature's most exquisite structures.

4. Stay in a room with a spider for a while.

5. Examine a spider with a magnifying lens.

6. Let a spider walk over your hand. (Now, give yourself a medal!)

Should all of these measures be of no help, find a good psychological counselor, because having such an extreme fear can cause you physical harm in the long run.

Advertisers play on this widespread irrational fear of the insect world. Nightly in the early 1980s, American television viewers watched a middle-aged couple quiver in terror because the husband had found "them" in the bathroom. After he set off an aerosol chemical, the couple collapsed in relief. "They" were gone. What the ad failed to say was that unless the neurotic pair changed whatever was making their bathroom attractive to "them" (most likely a leaky plumbing fixture), the insects would *always* return.

Another series of commercials had boxer Muhammed Ali urging us to get the "b-u-u-g-s" under control by fogging the air with a chemical. Cartoons of little crawlers marching into the mist and then obligingly falling belly-up emphasized the message. What the helpful heavyweight didn't tell viewers was that there are other, longer lasting ways of control. The insects he was talking about looked like cockroaches, and the champ failed to say that cockroaches learn to skirt most chemical poisons.

Observe How Children Relate to Other Little Creatures

Children usually have no insect hangups. The younger they are, the more fascinated they seem to be with all kinds of small wildlife. At age eight, Susie loved to catch honeybees in a milk bottle filled with purple clover and bring the buzzing bottle to friends as a gift. Six-year-old Mike would grin delightedly when a monarch butterfly, newly emerged from its cocoon, rested on his shoulder.

At age seven, Paul loved to keep "Hamsterdam," his pet hamster (what else?), nestled in his shirt pocket, its two black eyes like an extra pair of shiny shirt buttons. Five-year-old Sarah would spend hours combing through her family's garden for pillbugs. When she'd find one, she'd lift it gently to her palm, where it would curl up in self-defense. "Ooh, look!" she'd whisper, "It's going to sleep," and place it tenderly back in the grass. Yet some twenty years later, she regularly set off bug bombs in her apartment.

Do we adults infect children with our own half-understood fears of bugs? Or is it that, as highly urbanized as we have become, we fear the outdoors and its wildlife?

Home gardeners, perhaps uncomfortable with nature uncontrolled, apply a whole pharmacy of pesticides to their plants. In wiping out the leaf nibblers, though, they also destroy those insects' natural predators (not to mention polluting the ground water), and so become trapped on a chemical treadmill.

If an intense, gut-level dislike of insects is your problem, the procedures suggested in the chapters that follow can help you by holding down the number of them that you will encounter. But there are two other problems with which many people struggle. While they are ultimately beyond the scope of this book, they do deserve to be mentioned here. These are:

1. Imaginary insect infestations; and

2. "Cable mite dermatitis."

Let us take a brief look at each in turn.

Imaginary Insect Infestations

Some people are convinced that their bodies are riddled with invisible insects that cause them great discomfort. So real do these nonexistent creatures seem to the sufferers that they often persuade their relatives that the infestation is real. This condition is called *delusory parasitosis.*

"We [would] get half a dozen calls or letters a month," said James M. Stewart, formerly of the U.S. Centers for Disease Control and Prevention, "from people who are sure they're infested with one insect or another. One man thought that moth larvae were dropping on him and burrowing under his skin. We try to tell them what their problem may be without actually telling them that they're crazy."

To prove the existence of their tiny tormentors, people with this problem will bring to dermatologists or entomologists pieces of the animals' "bodies." These usually turn out to be skin scrapings, lint from their clothing, or products of the person's own oil glands. One deluded man, a successful playwright, placed his "vermin" on a piece of paper and made it jump. He ignored the fact that he had created static electricity by walking over carpet, according to Walter Ebeling, the UCLA entomologist who tried to help him.

Delusory parasitosis can be serious, and even potentially life-threatening. Researchers tell of a woman who boiled her family's clothing every night in an effort to dislodge "mites." She also made her children wash themselves with gasoline, turning them into walking Molotov cocktails. In the 1930s, entomologist Hugo Hartnack once tried to help a husky police officer who broke down and wept when describing his "infestation." Although the naturalist listened sympathetically, he could not help the man. A few days later, the police officer committed suicide in despair.

If someone close to you thinks that there are mites or insects on his or her body even though a dermatologist has said there are none, try to persuade the person to see a psychiatrist, because this is a psychological problem, not a physical one.

Cable Mite Dermatitis

A spinoff of this delusion about insects is a condition known as *cable mite dermatitis*. People who work in areas of low humidity and under stressful conditions sometimes develop a mysterious itch. They may be convinced that invisible organisms on the papers or telephones that they must handle on the job are causing the rash.

The power of suggestion is strong among groups of people who work together, and the cable mite problem can be a serious one for a company. In September 1980, Republic Airlines had to send eighteen of the twenty employees in its Los Angeles office home sick in one week. In fact, they were not ill because of the skin irritation that some of them had, but because of the pesticide that was being used to wipe out the *nonexistent* bugs.

In a later incident, described in the professional journal *Annals of Allergy*, workers in a physics laboratory developed dermatitis, and were positive that their rashes came from some microscopic animals in the lab. Luckily, before the foggers and sprays arrived, someone discovered that flaking rock-wool insulation in the ceiling was drifting into the ventilation system and causing the dry, itchy skin. In this case, some careful detective work and a slight modification of the building were the solution to a real dermatitis with a misunderstood cause. (It should be noted, though, that if rats or mice infest a building or wild birds nest in an attic, mite attacks on humans are a very real possibility.)

As with many of life's problems, having a sense of proportion is a basic element in controlling unwanted insects. If you see one confused, wandering ant, it doesn't mean that your home is being overrun. Nor does a nibbled leaf or two sound your garden's doom. Hold back the heavy artillery until you can tell whether you have a real infestation or just a brief encounter. If the ant turns out to be leading a battalion or if the nibbles start to spread from leaf to leaf, then you might consider taking some action. Keep in mind that overreacting may only make the situation worse, whereas calmly searching for the things that are attracting the insects, and utilizing a few safe controls such as those described in the following chapters, will probably clear out the invaders and prevent future assaults.

PART TWO

Pests of Food

3. Outsmarting the Cagey Cockroach

In the seventeenth century, any Danish sailor who caught 1,000 cockroaches aboard ship received a bottle of brandy. A hundred years later, Captain Bligh ordered the *Bounty*'s decks drenched with boiling water to kill the voracious insects that were devouring his cargo of breadfruit trees. In the nineteenth century, for every 300 seafaring roaches he trapped, a Japanese sailor won a day's shore leave.

The United States Navy sprayed 10,000 gallons of pesticide on its ships in 1978 in a futile effort to dislodge this pest. More recently, the American seamen have tried using sexual lures baited with substances called pheromones as roach traps. After a three-month test aboard a ship of the Atlantic fleet, the Navy admitted failure and gave up.

If powerful, efficient navies can't control cockroaches, what can the average householder do against these pests?

Plenty!

While not as exciting as shore leave, brandy, or sex, you can be rewarded with reliable cockroach control by tight maintenance of your home, cleanliness, and the strategic use of boric acid, hydramethylnon, or fipronil (the latter two are relative newcomers to the pest control arsenal) used with baits in small matchbook-sized traps. First, distasteful though it may be, you will need to learn a few facts about this pest.

TYPES OF COCKROACHES

There are a number of different types of cockroaches you may meet up with. For illustrations, see Figure 3.1.

The American Cockroach

One of the largest cockroaches, the average American cockroach is about $1\frac{1}{2}$ inches

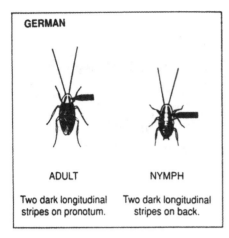

GERMAN

ADULT

NYMPH

Two dark longitudinal stripes on pronotum.

Two dark longitudinal stripes on back.

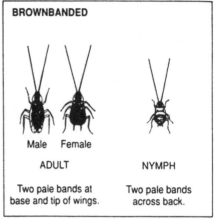

BROWNBANDED

Male   Female
ADULT

NYMPH

Two pale bands at base and tip of wings.

Two pale bands across back.

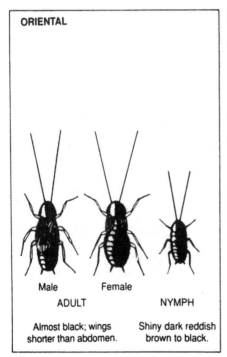

ORIENTAL

Male   Female
ADULT

NYMPH

Almost black; wings shorter than abdomen.

Shiny dark reddish brown to black.

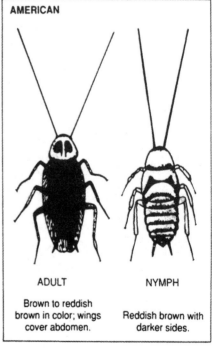

AMERICAN

ADULT

NYMPH

Brown to reddish brown in color; wings cover abdomen.

Reddish brown with darker sides.

Figure 3.1 Common Types of Cockroaches
If your Croton bugs, water bugs, shad bugs, palmetto bugs, black beetles, or Bombay canaries look like any of these insects, you have cockroaches. Most need a warm, dark, moist environment.

Source: Controlling Household Cockroaches. (Division of Agricultural Sciences, University of California, Leaflet 21035, 1978).

long. Reddish or dark-brown in color, American cockroaches prefer a warm environment, and their eggs need a high humidity level. They are often found in sewers and steam tunnels used to heat large housing developments, as well as in masonry storm drains. They may be attracted to outside lights at night.

The Brownbanded Cockroach

This pest has brown and yellowish bands across its wings and is usually about $\frac{1}{2}$ inch long. It prefers a warm, dry environment. Often found throughout the home, it lays its eggs in clusters on dark vertical surfaces such as those provided by bookcases, draperies, and closet walls, and inside dressers. This pest is often found high up in rooms, and populations can reach huge numbers.

The German Cockroach

The most common cockroach species found in homes around the world (and, curiously, called the "Swedish cockroach" in Germany), the German cockroach is light brown in color, is about $\frac{1}{2}$ inch long, and has two indistinct black bars on the head. It prefers warm, moist areas such as kitchens, bathrooms, and crawl spaces. These creatures are extremely fertile; one female can produce in her lifetime about 270 young, who may mate with each other and in turn can have 30,000 descendants in one year (some of which are usually cannibalized, but unfortunately this doesn't really lessen the problem).

The Oriental Cockroach

Blackish in color and about 1 inch long, the Oriental cockroach has the additional distinction of smelling bad. It is drawn to excrement and prefers damp, cooler areas such as basements, dense vegetation, water meter vaults, and floor drains, as well as sinks, bathtubs, refrigerators, washing machines, and water heaters.

WHAT COCKROACHES EAT

Cockroaches are primarily carbohydrate-eaters—that is, they mostly feed on things of vegetable origin. Since they also like meat, however, they are omnivorous, and will eat just about anything, including grease, sweets, paper, soap, cardboard, book bindings, ink, shoe polish, toothpaste, and dirty clothes. They have even been known to gnaw on the fingernails and eyelashes of sleeping sailors and infants. They are especially fond of beer.

True survivors, cockroaches can live for up to three months without food and thirty days without water. They react faster than the blink of an eye to a footstep or sudden light. Since cockroaches taste their food before eating it and learn to avoid chemically treated surfaces, most chemical pesticides do not give good long-term control.

THE HAZARDS OF COCKROACHES

Cockroaches themselves are as fastidious as house cats, forever cleaning themselves and polishing their sleek shells, but if one should crash your next dinner party, you could find your circle of friends abruptly shrunken. Few creatures arouse such loathing—and with good reason. They are known carriers of the following diseases:

• Boils and other skin infections

• *Escherichia coli* (*E. coli*) infection

• Gastroenteritis and other digestive disorders

• Hepatitis, an inflammation of the liver

• Plague

• Poliomyelitis (polio)

• Shigellosis

• *Staphylococcus* infection

• *Streptococcus* infection

• Toxoplasmosis, a parasitic infection that is usually harmless to healthy adult humans but that can be devastating to an unborn child or an immunosuppressed person of any age;

• Typhus.

Cockroaches have also been implicated in the spread of dysentery and typhoid fever. In poorer residential neighborhoods, where there is apt to be less control of these pests, their feces, carcasses, and sheddings cause high rates of childhood asthma.

Cockroaches can spread *Salmonella* bacteria, a common cause of food poisoning, which have survived on corn flakes and crackers for up to four years after the cockroaches have moved on. If you find even one of these vermin in any of your food stocks, you should throw that food out immediately.

Like all insects, cockroaches are extremely prolific but, unlike more seasonal species, they breed all year long, sometimes even without mating. Some years back, in tests of new pesticide formulas, 5,000 cockroaches that were set loose in a Raid research kitchen had multiplied to 16,298 when the room was fumigated one week later. While in their eggs, which are clustered in yellowish purselike cases, young cockroaches are immune to all insecticides. (See Figure 3.2.).

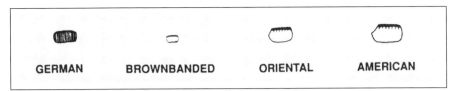

Figure 3.2 Cockroach Egg Capsules
Source: Controlling Household Cockroaches (Division of Agricultural Sciences, University of California, Leaflet 21035, 1978).

You think you've got roach problems? Consider this: It was reported in *Pest Control Technology* in 2002 that a company had recently vacuumed up *twenty pounds* of cockroaches from a Manhattan apartment that had been vacated shortly before by a recluse.

Despite their well-deserved bad reputation, however, cockroaches are actually of some use to us. Researchers have used them in investigating cancer, heart disease, nutrition, and the effects of space travel. They are hardy, reproduce quickly, and have many biological systems that are very similar to ours, so they make great subjects for study. They are also vital for our forests, where they turn dead vegetation into valuable soil nutrients.

HOW COCKROACHES MAKE THEMSELVES AT HOME

Chances are that you are reading this book because you have been sharing quarters with cockroaches. So common are these pests—and so anxious are we to get rid of them—that they have been called the pest control operator's bread and butter. Few insects are so clever at making themselves comfortable in our homes. They have been found nesting in telephones, televisions, radios, refrigerators, electric clocks—even a flute. People move out of their apartments because they cannot cope with a cockroach infestation. They are very common in substandard, dilapidated, and crowded homes and apartments.

If you see even one cockroach in the daytime, you can be certain that you are hosting many more. Most cockroaches hide from light, and expose themselves to it only if they are crowded out of their hiding places.

SHUT THEM OUT

The critical step in getting rid of cockroaches is to change whatever makes your home attractive to them. Cockroaches are tropical and seek warmth and moisture. To the extent that your home provides these conditions, it will draw the insect.

A thorough cleaning of all infested rooms—and even those that *seem* to be cockroach-free—is a good start. Use soap and water on all washable surfaces, and your vacuum cleaner—if possible, one equipped with triple filters or a high-efficiency particulate absorber (HEPA) filter—on the others. Following are additional recommendations that apply to specific rooms in the home.

The Kitchen

❑ Clean the area around your water heater, especially if it is housed in a separate closet.

❑ Move the refrigerator and stove out into the room (you may need to recruit a muscular helper for this job), and scrub the walls and floor behind them with an industrial-strength cleaner. When the floor is dry, dust it *lightly* with *technical* boric acid powder, which is available in hardware and home-repair stores. (A detailed discussion of boric acid, which is a stomach poison, can be found on pages 47 through 50.)

❑ If the wallpaper behind your appliances has come loose, glue it down securely. Loosened paper makes a fine hideout for all light-fleeing insects.

❑ While the refrigerator is pulled away from the wall, check the outside of the box for small smears or threadlike droppings. These, along with an unpleasant odor, are a sure sign of cockroaches in the insulation, a favorite and hard-to-reach harborage. Thousands can live in one refrigerator. If you do find droppings, have the box fumigated—or get a new one.

❑ When shopping for a used refrigerator, *beware* is the word if the seller is a private party. You could be buying a cockroach infestation along with that bargain box. You would be better off finding a reputable used-appliance dealer, who will have had all the refrigerators they sell fumigated as a matter of course.

❑ Make sure your kitchen cupboards and drawers are very clean. These often have little caches of food crumbs that can keep any food pest fed indefinitely.

❑ Take a look under your kitchen sink, which could tell you why you're having a problem. Do you keep bags or boxes of pet food there, along with your damp dishrag? If so, you are duplicating a roach-breeding arrangement maintained at one time by the famed San Diego Zoo to raise food for its exotic birds. The zoo's birdkeepers housed large colonies of the insects in twenty-gallon garbage cans filled with rolls of corrugated cardboard surrounding a jar of water containing a cotton wick. The whole assemblage was generously sprinkled with dog kibble. So nuke your dishrag in the microwave or run it through your dishwasher, and dry it thoroughly before putting it back. Store pet food in a tightly covered can.

❑ If you have been finding cockroaches in the sinks in the morning, they may be climbing up from the sewers. *Do not* pour a pesticide down the drain. This is illegal and also contaminates drinking water and harms the environment everywhere. These chemicals can also deactivate septic tanks. Instead, call your local sanitation department or public health authorities to come and treat your sewer lines. Until they do, close all your drains at night and stuff the overflow hole with a wad of steel wool wrapped in a small plastic bag.

The Bathroom

In the bathroom, water, rather than food, is apt to be the primary lure. Check for the following:

❏ Be aware that danger spots are wherever moisture gathers—the base of the toilet, the angle between countertop and backsplash, and anywhere else where dampness is common. Dry these up and keep them dry.

❏ If you have a vanity cabinet, check for water around the pipes underneath.

❏ After using a rubber shower mat, hang it up to dry. Its rows of suction cups can give a cockroach a custom-built hiding place.

Other Hideouts

Although cockroaches prefer kitchens and bathrooms, they can easily fan out to other parts of the house. Following are some suggestions for foiling them:

❏ Inspect and clean thoroughly all closets, television sets, radios, and electric clocks.

❏ Vacuum any upholstery near the TV to pick up crumbs from snacks eaten there. (If you can, confine your family's snacking to the kitchen or dining room.)

❏ Keep computer areas free of food. Insect parts can ruin a computer.

❏ Empty all bookcases and vacuum the shelves. If they are washable, scrub them. Shake out each book and flip its pages wide open. A couple of cockroaches—along with some silverfish (nature's own paper-shredders)—may tumble out.

❏ Cockroaches like the binding glue used in books. Brownbanded cockroaches, which are phenomenal reproducers, attach their egg cases to large vertical surfaces.

❏ Vacuum the folds in your window draperies. These also are dark vertical surfaces that may harbor cockroaches. Periodically taking them down and either washing them or having them dry-cleaned may help as well.

❏ Vacuum and/or wash corners under tables and chairs, which are favorite hiding places for brownbanded cockroaches.

❏ Dispose of the vacuumed-up material by burying it deep in the ground or placing it in a tightly closed garbage can set out in direct sun. Either of these methods kills the vermin's young.

❑ Check your garage, basement, and/or attic, and get rid of those piles of old magazines and newspapers, cardboard boxes, and discarded clothes. Any of these can support large colonies of cockroaches. Corrugated boxes are especially bad, because the corrugations easily shelter the tiny cockroach nymphs.

❑ Store mops and brooms upside-down so that they will dry quickly and cockroaches will not hide in them.

❑ Take care of those long-delayed small repair jobs. That dripping faucet and seeping toilet are reservoirs for insects. A new washer may take care of the faucet, but the toilet could need a plumber's services.

❑ Fill all cracks in the walls and woodwork, including crevices between baseboards and floors, any spaces between cupboard walls and shelves, and cracks around drainpipes and sinks. A caulking gun works well for this. For cracks that are hair-fine, you can use latex paint to seal them off. Before deciding to ignore a crack, remember that a cockroach is so flat that it can hide in a space only $\frac{1}{16}$ inch high. If you are short of time, stuffing these hideouts with steel wool is a good stopgap measure. Incidentally, these are all excellent controls for carpet beetles and silverfish, too.

❑ If your sink's splashboard and/or its counter come loose, glue them down securely. Both are often the main sites of infestations. Be sure to caulk all joints.

❑ Repair cracks in stucco or cement with the right patching compound to close up even the smallest opening.

❑ Minimize dampness, wherever it is.

❑ Seal any gaps around water or gas pipes coming into the house. These are virtually cockroach expressways.

❑ Put several inches of gravel into your water meter vault. Its damp cavity is a perfect roach incubator.

❑ Pull up dense patches of ground cover, especially Algerian ivy and other broadleaved plants. Thin out other plants to eliminate dampness.

❑ Clean up any debris that may be present.

❑ Move stacks of lumber and cordwood away from the house.

❑ Finally, some useful "Don'ts":

• *Don't* save grocery bags. They are frequently infested, and even if they are not, they can provide harborage for cockoaches that come in by other routes.

- *Don't* buy beverage cartons that have spilled syrup or malt. You could be buying cockroaches. One soft-drink bottle that was returned for deposit was found to hold 200 young vermin.

- *Don't* save corrugated cardboard boxes. Cockroaches love them. If you are collecting boxes because you are planning to move, look them over carefully to make sure you don't carry a verminous family to your new home along with your pots and pans. Grocery cartons are more apt to be infested than are liquor or book cartons.

- *Don't* buy used furniture without inspecting each piece with a wary eye. That mellowed dresser or bed could house some unwanted occupants.

Apartments

Does your apartment sparkle, but you still spot occasional scurrying brown crawlers when you turn on the kitchen or bathroom light at night? The insects could be migrating into your place from a neighbor's. To stop the intruders:

❏ Nail window screening over the heating vents between your quarters and those of other tenants.

❏ Weather-strip your door to the common hall.

❏ Be especially watchful if your building has an incinerator or a compactor. If these malfunction, or the area around them isn't kept clean, they attract vermin. Incinerator temperatures should range from 1400°F to 2000°F. The apartments' management should make sure that Dumpsters are emptied and cleaned often, and that their lids are kept closed.

STARVE THEM OUT

Cockroaches in your kitchen are a loud signal to you to change the ways in which you manage food and water. The following questions and answers will tell you how.

❏ Do you store your staples, once you have opened the original packages, in tightly closed, impenetrable containers? Clean pickle jars, tea canisters, and tin cracker boxes are all suitable for storing dry foods; you don't need decorated kitchenware for this job.

❏ Do you always clean up spills and crumbs promptly and sweep the kitchen floor daily? From an insect's eye-view, a few crumbs are a feast.

❏ Do you put your kitchen garbage in a plastic bag that is kept closed with a twist tie to prevent odors during the day? If not, your garbage could be sending out a pest-drawing aroma.

❑ Do you put your garbage out every night? That's when roaches are most active.

❑ Do you leave snacks out for your pet? Kibble is as delicious to vermin as it is to dogs and cats. Remove the animal's food right after feeding time.

❑ Do you leave your pet's water dish out overnight? If your pet needs a nighttime drink, set its water dish within a larger pan that is filled with detergent suds.

❑ When do you wash your dinner dishes? If you absolutely *must* leave them unwashed for a few hours (a practice definitely not recommended), submerge them in water and detergent.

❑ Are there little reservoirs around your home? The most unexpected watering holes attract these insects. Cover fish tanks, flower vases, and not-quite-empty soft-drink bottles; refrigerate overripe fruits and vegetables; and dry up the catch pan located underneath the bottom compartment of your refrigerator.

WIPE THEM OUT

Now that your home is no longer fit for cockroaches and you are no longer encouraging them outdoors, you can eliminate those left inside. Before grabbing for the fogger or spray, though, remember that after repeated use, most chemical pesticides have no effect on cockroaches. They learn to avoid them, or they become biologically resistant to them. Therefore, alternative methods—like traps—may be your best choice.

Traps

If you have young children and do not want to use boric acid (see page 47), here are some traps that have proven effective:

❑ The British trap. Twelve of these fragrant lures have been reported to have netted over 8,000 vermin in the United Kingdom during a three-month period. Of course, British cockroaches would like beer, but their cousins around the world also seem to find it tasty. To make one, follow these steps:

1. Wrap masking tape (to give your prey a foothold) around the outside of a good-sized empty jam jar.

2. Half-fill the jar with beer, a few banana slices or pieces of banana peel, and a drop or two of anise extract. Boiled raisins are also good bait.

3. Smear a thin band of petroleum jelly one or two inches wide just inside the rim so that your victims can't climb out.

4. Push the trap up against a wall or other vertical surface. Cockroaches tend to follow intersections, such as those where walls and floors meet.

❏ The Texas trap:

1. Paint a wide-mouthed pint jar flat black.

2. Smear petroleum jelly inside the rim as described on page 42.

3. Add generous amounts of pet kibble or pieces of apple, potato, or banana peel.

❏ The simplest trap of all. In the mid-1930s, pest control expert Hugo Hartnack recommended soaking a rag in beer and leaving it overnight on the floor where you have seen cockroaches. In the morning, you just step on the drunken bugs.

❏ Commercial traps. Roach Motels and other such baited boxes attract German cockroaches, but fewer of the other species. Also, they may draw ants, which show up to consume the dead vermin. Professional pest control operators may use them to judge how severe an infestation is, but they don't rely on these devices to wipe it out. Female cockroaches carrying egg cases do not feed or move around much, so baits don't catch many of these. However, since cockroach droppings contain natural chemicals that attract other cockroaches, placing baits near such signs of infestation can be effective.

Whichever traps you use, set them upright against corners of the room or under a sink, wherever you've seen cockroaches. To catch German cockroaches—the most common variety—traps must be inside; those placed outside will catch other species. (See Figure 3.3 for trap placement.) Drown your catch in a bucket of hot, sudsy water. Of course, you should always wash your hands after handling traps.

No type of trap or bait box will work very well if the insects can easily find other food. Good housekeeping makes hungry cockroaches easier prey. If you leave pet food out overnight and set a trap or bait station nearby, you will fool no one but yourself. A retired college professor and his wife did just that, and couldn't figure out why the cockroaches went for the moist cat food and ignored the trap.

Chickens: An Efficient Cockroach Patrol

If you live in a rural or semirural area, consider keeping a few chickens. Apart from having a ready source of eggs and poultry, you will have many fewer vermin around your home. As an old Spanish proverb says, "When arguing with a chicken, the cockroach is always wrong." It's worth a try.

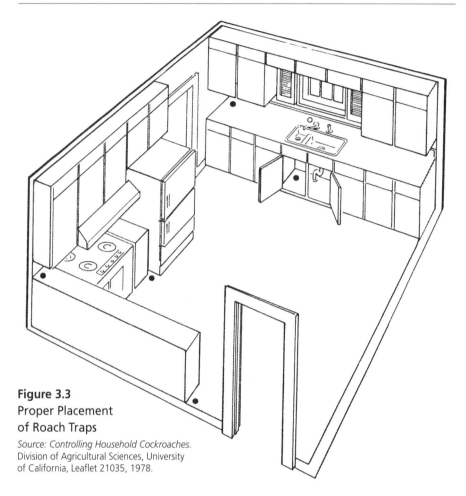

Figure 3.3
Proper Placement
of Roach Traps
Source: Controlling Household Cockroaches.
Division of Agricultural Sciences, University
of California, Leaflet 21035, 1978.

Temperature: Hot, Cold, or Just Right?

Temperature can attract the pests or drive them out. Setting your thermostat at a steady 81°F is a welcome mat for cockroaches, as well as for many other insects. To discourage them, it is best to keep your home on the cool side—the cooler the better.

The British Museum (Natural History) reports that a temperature of 130°F has been used successfully in steam-heated apartment buildings to kill cockroaches that are established in the heating systems. Radiators were closed to trap the insects, and boilers loaded to capacity. According to the Pasadena, California, fire department, having your heat turned up to such a sweltering temperature for several hours poses no fire hazard—although if you don't close off the hot-air grilles of radiators, you do run the risk of damaging furniture, musical instruments, computers, and other household items. Also, if the wiring in your home is made of aluminum, it may not be able to handle the high current needed to run your furnace at top capacity, so doing this could blow your circuit breakers. If you decide to try this, be sure to monitor your home closely.

At the other end of the thermometer, cockroaches die at 23°F. If you live where the winters are cold, and circumstances permit, leave your windows open all day in freezing weather to wipe out an infestation. (This will also help to eliminate silverfish and clothes moths.)

Silica Gel

If traps and temperature extremes fail, but you are not yet ready to use boric acid, you might want to try silica gel sorptive dust. Applied like boric acid and in a powder form, it destroys the insect's waxy coat, and the animal shrivels and dies. It is best for use in attics, wall voids, and other enclosed places. Silica gel, however, is very repellent, and the pests may soon learn to avoid it. Also, it not only costs more than boric acid, but is highly irritating to the lungs and tends to float all over the house, clinging to curtains and absorbing wax from furniture. However, despite the skull and crossbones on the package, it is not actually toxic.

FUTURE CONTROLS

Scientists at Yale University have synthesized a sex lure that sends male cockroaches into a frenzy, drawing them into a trap. Unfortunately, the lure drives only the American cockroach crazy. It leaves the German cockroach cold, and this is the species most commonly found in our homes. And while researchers are optimistic about the product, you should not expect to find it on store shelves soon.

Folk wisdom has long held that catnip will repel roaches. In 1999, two scientists at Iowa State University reported evidence that confirmed this belief. They pinpointed nepetalactone, one of the plant's components, as the substance that drives the insect off. There is not yet any commercial cockroach-control product based on catnip, however (and I personally wouldn't spread catnip outside my house since I wouldn't want to attract any more neighborhood cats than already show up), but the judicious use of catnip in the home might prove effective.

In the 1990s, Heather Wren, an entomologist at Virginia Polytechnic Institute and State University (Virginia Tech), developed a "cockroach birth-control pill" that prevents cockroaches from reproducing. The compound is now being marketed by Cleary Chemical, Inc. as the Ecologix Roach Terminal. It is one of a group that this scientist has developed that stop an insect's ability to store nutrients.

Without adequate nutrition, the insect dies of exhaustion. Says Dr. Wren, "It's not fast, but it is effective, and there is no residual poison to enter the food chain such as if a bird or other animal eats the dead insect."

Another antiroach strategy, developed in 1985 by the U.S. Department of Agriculture, is a repellent so poweful that even starving cockroaches will not enter a treated area. Potential applications include protection of electronic equipment and of breeding places, like cardboard containers. This product also may not be on the market for several years.

Keep in mind, though, that whatever type of pest controls or repellents you may choose to use, you should not slacken your home maintenance routine. As the United States Navy has found, sex lures alone may not do the job.

CHECKLIST FOR COCKROACH CONTROL

❑ Thoroughly clean bookshelves and closets; behind the refrigerator and stove; and inside televisions, electric clocks, and radios.

❑ Caulk all cracks in interior and exterior walls. Close up openings around pipes.

❑ Repair all plumbing leaks.

❑ Secure splashboards behind sinks and countertops.

❑ Check the inside of your water meter.

❑ Pull up ivy or other broad-leaved ground covers.

❑ Clean up lumber and any other debris near your house. Move any wood, lumber, or firewood away from the house.

❑ Check grocery bags, cartons, and soft-drink cartons before bringing them inside.

❑ Store all food, including pet food, in tightly closed glass, metal, or plastic containers.

❑ Never leave dirty dishes in the sink for any length of time.

"Help! A roach!" most of us feel like yelling when we find one of these vermin. But if you can calm down long enough to make sure your home is not the kind of habitat cockroaches need, you can get rid of them—and of the chemicals commonly sold to kill them—for good.

To reach zero cockroach population in your home, you will need to use several different tactics at once. However, careful housekeeping must always be at the head of the list. *Good sanitation is always your strongest weapon in the battle against the cagey cockroach.*

Boric Acid, Hydramethylnon, and Fipronil: Exceptional Means to Eliminate an Exceptional Pest

Because an infestation of cockroaches is so hard to eliminate, and even the most meticulous of housekeepers has occasional lapses that favor this pest, when all of the measures suggested in this chapter do not succeed in completely clearing out a colony, I do recommend the use of three toxic substances: boric acid, hydramethylnon, and/or fipronil. Boric acid has been used in cockroach control for generations, whereas hydramethylnon and fipronil are relative newcomers. These are the safest chemical controls we now have. Set in out-of-the-way places and, in the case of hydramethylnon and fipronil, packaged in child-resistant bait stations, these chemicals provide the least toxic long-term solutions currently available. As with all pest control chemicals, hydramethylnon (an ingredient present in products sold under the brand names AC 217,300, Amdro, Combat, Maxforce, and Wipeout, among others) and fipronil (in Frontline, Combat, MaxForce, and other products), and boric acid should all be used with great caution.

BORIC ACID: THE PROVEN KILLER

For the long haul, the safest, most potent cockroach-killer is technical boric acid. It is sold commercially in various brand-name products, such as Roach Pruf. These agents are formulated specifically for pest control.

Known for generations as a reliable medical antiseptic, boric acid powder seems to work as an insecticide in two ways. First, the powder—a stomach poison—is picked up on the animal's feet, later to be swallowed when the pest preens itself. Second, the powder penetrates the cockroach's outer covering, leaving it vulnerable to dehydration.

Although cockroaches have learned to avoid chemical pesticides that they can smell, and have developed immunity to others, they have never learned to avoid boric acid or become resistant to it. Boric acid has been used successfully against cockroaches since the beginning of the twentieth century.

Use Only Technical Boric Acid

If you decide to try boric acid, you should know, first of all, that you must use *technical* boric acid. It is illegal—as well as unwise—to use *medicinal* boric acid against roaches. Medicinal boric acid (which is used as a surface antiseptic and antifungal, and is never to be taken internally) is very easily confused with table sugar or salt. Swallowing even a tablespoonful of it can kill a small child or give an adult digestive upset. Boric acid that is sold for pest control is usually tinted for easy identification. Further, it usually contains an additive to prevent clumping and has its natural electrostatic charge augmented so that more of the powder clings to insects' bodies. Thus, a powdered cockroach carries the poison back to its lair, where others pick it up.

If sprinkled in out-of-the-way corners, technical boric acid does no harm unless ingested, and since it does not interact with air, it continues to kill any cockroach that comes into contact with it for as long as it is in place. Boric acid made specifically for insect control is available in hardware and builders' supply stores.

How to Use Boric Acid

The best way to use technical boric acid is to place small amounts in secluded areas so that it forms a *light* layer on all surfaces. About one to two pounds will take care of an average apartment; about two to four pounds will clear out a moderate-sized house.

Apply a film of the powder along kitchen and bathroom baseboards, behind and under stoves and refrigerators, under your kitchen sink, at the back of the bathroom vanity cabinets, and behind toilets. (See Figure 3.4.) Do not pile it up. You can also blow the boric acid powder into the razor blade slots of a medicine cabinet to disperse it in the wall void behind the bathroom.

Do not sprinkle boric acid *anywhere* you store food. To penetrate the empty spaces under your kitchen cabinets—a great cockroach lair—drill a $1/8$- to $1/4$-inch hole every five or six inches at the top of the kick panel or in the bottom shelf of each floor cabinet, and blow in the boric acid with a powder blower, a bulb duster, or a squeeze bottle such as those used for ketchup or mustard. Also, be sure to label the container clearly and permanently. Unlike the medicinal form, electrostatically treated boric acid does not clump.

At night, look for other hiding places. Enter your darkened kitchen and bathroom and shine a narrow-beamed flashlight on likely lairs. The cockroaches will run from the light, but not before you have had a chance to see where they're going. Then treat those areas.

Boric acid works more slowly than other anti-cockroach products. A day or two after applying it, you might notice one or two small corpses; or you may see more of the live pests than you had before as the dying insects wander out

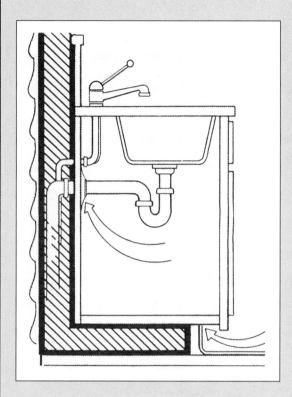

Figure 3.4
Application of Boric Acid Powder Around a Kitchen Sink

The dark lines indicate a thin layer of the powder. Access to these areas may be found at corners through small cracks and at plumbing and electrical openings in the wall.

Source: Controlling Household Cockroaches (Division of Agricultural Sciences, University of California, Leaflet 21035, 1978).

of the walls. Satisfactory control takes anywhere from two to ten days, by which time the number you see should drop sharply. If you keep up your good housekeeping and home repairs, months or years may go by before any more show up. The fatal powder just keeps on working. Of course, if you remove the boric acid while cleaning, you will need to reapply it.

A *tip:* If you are building or buying a new house, or are remodeling your kitchen and/or bathroom, this is an ideal time to put in a very effective cockroach-control measure. It is easy to dust attic, wall voids, dropped ceilings, and spaces under cabinets and built-in appliances before they are enclosed. Before the refrigerator is set in place and the cabinets are installed, sprinkle the floor where they will be set with a light dusting of technical boric acid. This treatment can be effective for years. I did this in my own kitchen during a remodeling in 1997 and have not seen a cockroach since. (Unfortunately, I can't say the same for ants or crickets, but I still feel I'm ahead in this battle.)

A *note of caution:* Because boric acid powder is very slippery, you should avoid sprinkling it where people walk. It is also a stomach poison, so if you are going to be in contact with it for an extended period while placing it around the house, wear gloves and a particle mask. Also, plants are easily killed by the powder, whether by direct contact or through the soil. Needless to say, it must be kept away from children and pets.

HYDRAMETHYLNON AND FIPRONIL

In 1987, the insecticide hydramethylnon appeared on supermarket shelves. A few years later, another chemical, fipronil, began to be used for cockroach control. Packaged in child-resistant bait stations (a big plus in any pest control product), these chemicals keep cockroaches from digesting their food so that they starve to death. According to Donald A. Reierson, one of the scientists at the University of California, Riverside, who investigated the insecticide, hydramethylnon is "twice as safe" as boric acid. And besides, the bait stations disperse the chemical in tiny amounts, much smaller than those needed for an effective treatment with boric acid. So any child who manages to pry open the small black containers would have to swallow the contents of forty or fifty of them to get sick.

Since it cannot be blown about or scattered except in its secure containers, bait products containing hydramethylnon, such as Combat, are safe to use in utensil drawers and on shelves where food is stored—places where you should never sprinkle boric acid. And if the places that are harboring the pests are inaccessible—for instance, areas under a tiled counter top—the bait stations will eventually clean them out.

As with all pest control methods, hydramethylnon and fipronil are not the whole answer. Many are not laced with any type of attractant, so it is hunger and the lack of other food that catch their victims. These traps work only in clean quarters. Also, with boric acid, one cockroach in effect serves as executioner for its fellows by carrying the powder back to the hideout they share. Hydramethylnon and fipronil work only on the particular insects that eat them. Manufacturers claim that if you use products containing these chemicals according to their directions, you should notice a decrease in cockroaches within a week.

It is not known whether any cockroaches have developed immunity to the bait stations, but some apparently have recovered from their severe stomachaches and now avoid the bait. Thus, bait stations are just one more tool—albeit a good one—in roach control.

4. Critters in the Crackers, Pests in the Pantry

A Pennsylvania homemaker is preparing supper and decides to serve her family some fresh cornbread, a treat she hasn't baked in months. Taking an open box from the cupboard, she notices that some of the cornmeal is sticking together in long threads. As she shakes the grains into her measuring cup, out flies a dark-winged moth. Unnerved, the woman throws the box of cornmeal into the garbage can under the sink.

Across the continent, in the state of Washington, a man who is fixing breakfast pours cracked wheat into boiling water and goes off to set the table. Uncovering the pot a few moments later, he finds some tiny brown insects with toothlike projections behind their heads, floating dead on top of the thickened gruel. He shrugs, skims off the little corpses, and serves his family the hot cereal.

These people have just met the two most common pests of the home pantry: an Indianmeal moth and a sawtoothed grain beetle. Each of these individuals will probably find more of these insects in the weeks ahead, and most likely in foods other than cereals.

Along with rats and mice, insect pests that infest stored food destroy more than $1 billion worth of food in the United States every year. In some developing countries, the loss tops 40 percent of total food products.

Besides grains, pantry insects attack legumes, pasta, dried fruits, cheese, nuts, and spices—even chocolate. One whole group, called keratin eaters, likes meats and, by extension, leather and furs, as well as fabrics, books, and wood. Another tough little character, the drugstore beetle, can gnaw through aluminum, asphalt, and lead to reach its food supply.

TYPES OF PANTRY PESTS

There are many species of pantry pests, some more common and some more destructive than others. All, however, can be grouped into five general categories: beetles; their long-snouted relatives, weevils; moths; mites; and cheese skippers.

Left unchecked, all of these arthropods can spread through a house, climbing up the walls and thronging around windows. Even if they are confined to the kitchen, they make most homemakers nervous. Figure 4.1 pictures the most common wildlife invaders of the home pantry.

Beetles

The larvae of these insects tend to feed on flour, dried fruits, cigarettes, prepared cereals, pastry and biscuit mixes, dried soups, herbs and spices, and cured meats.

Beetles come in a nearly endless variety. Following are some of those you are most likely to encounter.

Cadelles

Destructive aboard ships and in grain mills, cadelles are also a frequent pantry nuisance. They gnaw through sacks and cartons, and like to lay their eggs under loosened carton flaps. Their habit of burrowing into woodwork creates hiding places for other pests. They are shiny, black, and flat; about 3/8 inch long, with the head and thorax distinct from the rest of the body. A female can lay as many as 1,300 eggs in her two-month-long lifetime, usually in protected places, such as cracks.

Cadelles are most likely to infest cereals, nuts, potatoes, flour, fruits, and whole grains. Signs of infestation include the presence of adult insects or larvae (which are quite large—1/2 inch or longer).

Carpet Beetles

Are you surprised to find these notorious fabric-destroyers listed among pantry pests? Don't be. Carpet beetles frequently attack stored foods. In nature, they are primarily scavengers that feed on animal remains. (Keep an eye on any Egyptian mummies you happen to have around the house, for these beetles will devour them!) The adults visit flowers, where they feed on pollen; their young do the damage in our homes. Foods contaminated by carpet beetles can cause severe digestive upsets, and the pests themselves can trigger allergies.

Carpet beetles can go without food for as long as a year—a great convenience for a creature that must wait for a stray cadaver to show up before it can get a meal. A dead rodent within a building's walls will attract them. They are drawn to light and may throng around windows.

The appearance of carpet beetles varies from species to species. Their length can be anywhere from 1/8 inch to nearly 1/2 inch. They can be glossy black or have markings of yellow, brown, and white. Some are actually quite beautiful. The larvae are bristly and carrot-shaped. (See page 53.)

Carpet beetles favor dried milk, cayenne pepper, legumes, seeds, corn, wheat, rice, and dry pet food. The adults also feed on pollen. Signs of infestation include

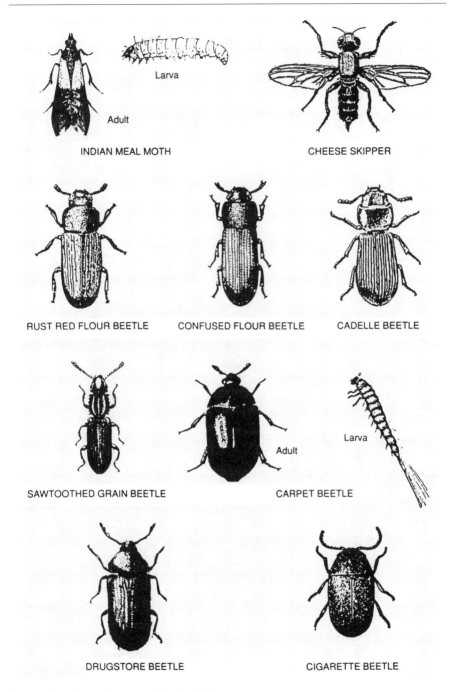

Figure 4.1 Some Common Pantry Pests

Source: Common Pantry Pests and Their Control (Berkeley, CA: Division of Agricultural Sciences. University of California. Leaflet 2711, 1982).

adult beetles, larvae, cast-off larval skins, and fecal pellets. The larvae are active and may be found anywhere in the house. In heavy infestations, adults are found swarming at windows.

Cigarette Beetles

This insect is especially fond of tobacco products, but don't think you're safe if you are a nonsmoker. Cigarette beetles attack a great variety of foods as well.

These insects are about $1/10$ inch long, roundish, reddish-brown, and covered with short, fine hair. Their low-hanging heads give them a humped appearance. The larvae are small, wiggly, and cream-colored.

Cigarette beetles may infest herbs, spices, grains, dried fruits, meats, fish, vegetables, nuts, and coffee beans. Also leather, insecticides containing pyrethrum (this is some tough cookie!), furniture stuffing, and bookbinder's paste. Paprika and dog food are most often infested in the home.

Signs of infestation include the presence of adult beetles and larvae. There may be tiny holes in food containers, as well as webbing in tight places.

Drugstore Beetles

One of the toughest pantry pests, the drugstore beetle eats everything except cast iron. It also has the eerie ability to thrive on poisons like aconite, belladonna, and strychnine.

About $1/10$ inch long, these bugs are light brown, covered with silky down, and have low-hanging heads. They favor any foods, spices, bird seed, pet foods, and pastry mixes, as well as drugs, books, tin, and aluminum foil. Signs of infestation include the presence of adult beetles and larvae; cocoons covered with bits of food; webbing in tight places; and tiny holes in containers.

Flour Beetles

Along with the sawtoothed grain beetle and the Indianmeal moth, these are the most common pests in markets and the home pantry. Clever humans have actually found some practical use for them. Because they cannot climb glass, they make docile laboratory animals. And a thousand of them, along with some tiny wasps, were the first earthlings rocketed into space.

Flour beetles are about $1/8$ inch long, reddish-brown in color, and have a flat, oval shape. The foods they tend to infest include grains, legumes, shelled nuts, dried fruits, chocolate, spices, snuff (does anyone take a pinch of snuff these days?), and cayenne pepper. They can also ruin museum and herbarium specimens. Signs of infestation include clumps of food and/or insect eggs clinging to sides of containers, tiny holes in packages, webbing in tight places, or adult insects and their droppings. The adults give off a foul odor as well.

Sawtoothed Grain Beetles

Sometimes mistakenly called sawtoothed weevils, these tiny beetles resemble miniature double-edged saws. They are lively little cusses whose flat shape lets them penetrate sealed packages. They cannot damage whole grain, but once grain is milled or otherwise broken, they mount their assault. Their larvae are nearly invisible, so you may eat them without knowing it.

Sawtoothed grain beetles are bright brown in color and about $1/10$ inch long. They have six little "teeth" on each side of the thorax. They are drawn to cereals and other grains, spices, tobacco, meats, candies, and dried fruits. They are especially fond of raisins. Signs of infestation include the presence of adult beetles, webbing in tight places, and tiny holes in packages of foods and spices. To check for them, shake out the contents of one spice container onto a sheet of white paper, and watch for any movement. A magnifying glass can help you locate beetle parts. The beetles sometimes are found wandering around on floors, in silverware drawers, or in kitchen appliances. They can also bite, according to entomologist Carol Sutherland of New Mexico State University's College of Agriculture and Home Economics. I've had these in my kitchen, but evidently they didn't find me too tasty because I've never been bitten.

Spider Beetles

This category of beetle, which gets its name from its long, delicate legs, can infest a vast range of items, including stored foods, only some of which are listed below. Because they can survive very low temperatures, refrigeration does little to control them.

Depending on the particular species, these insects can be from less than $1/16$ to nearly $1/4$ inch long. Their color ranges from red to various shades of brown, black, or yellow.

Foods and other items that may be infested by spider beetles include almonds, cocoa, corn, dried soups and fruits, rye, spices, fish food, and raisins, as well as casein (a protein in milk and other dairy products), furs, feathers, silk, linen, packaging material, wood, lead, drugs, and excrement. Signs of infestation include the presence of adult insects and/or larvae, as well as food clumped around cocoons.

Weevils

The larvae of these long-snouted relatives of beetles go for rice, unmilled wheat, and dried peas and beans—all foods with hard outer coverings. They are extremely destructive and, like all insects, phenomenally fertile. As many as twenty-eight weevils have been known to develop in one bean.

Granary Weevils and Rice Weevils

These pests may be found anywhere in the house. Their long, hard snouts enable them to bore through hard substances. In cooler climates, the granary weevil is more apt to invade your pantry; in warmer regions, the invader will be the rice weevil. One key difference between the two species is that rice weevils can fly; granary weevils cannot.

Both granary and rice weevils are brownish or blackish in color and about ⅛ inch long. They have long, blunt-ended snouts. The larvae are half-moon-shaped and legless.

Granary and rice weevils tend to infest whole grains, pasta, beans, nuts, cereal products, grapes, apples, and pears. Signs of infestation include the presence of adult insects and/or larvae, as well as holes bored in foods.

Bean Weevils

Not actually true weevils (since they lack the characteristic long snout), these insects have been major pests throughout recorded history. They have been found in beans in ancient Inca graves, and are mentioned in the Talmud.

They are velvety gray or brown with long black and white markings. About ⅛ inch long, they have a chunky shape, with the rear wider than the head. Susceptible foods are beans, peas, and lentils. Signs of infestation include the presence of adult beetles as well as holes in legumes.

Flour and Grain Moths

Although adult moths are harmless, their larvae devour nuts, dried fruits, cornmeal, and many other grains. Major pests in grain mills, moths can also do a good deal of damage in the home. Their larvae spin cocoons in our food and otherwise contaminate it. They are extremely destructive. Perhaps the most common is the Indianmeal moth.

On first spotting one of these moths, you may wonder what a clothes moth is doing among the groceries. Don't be fooled. A clothes moth is a drab buff color, whereas a meal moth is more handsome, with either coppery or dark gray bands striping pale gray wings.

The adults have a wingspread of about ¾ to ½ inch. The larvae are about ½ inch long. Their color varies from dirty-white to pinkish or greenish hues.

Flour and grain moths are drawn to bran, biscuits, dog food, nuts, seeds (including bird seed), chocolate, crackers, powdered milk, candy, red peppers, dehydrated vegetables, and dried fruit. If you find "white worms" in dried fruit, they are probably grain moth larvae. The larvae can also be found in other parts of the house.

Signs of infestation include webbing in susceptible foods. One species of grain moth, which infests only whole grains, imparts a sickening odor and taste.

Mites

These tiny creatures like just about all damp foods that other pantry pests eat, plus a number of other foods, as well.

Like their relatives, ticks and scorpions, mites spell trouble for humans. They can carry tapeworms and cause kidney and skin problems. Grocer's itch, the common name for a severe form of dermatitis (skin inflammation), is caused by mites.

Mites spread very quickly from bird and rodent nests. You can easily pick them up and, unaware, carry them home on your shoes and clothing. They are almost microscopic in size and translucent. They have sparse body hair and move slowly away from light.

Mites may infest dried fruits, flour, meat, cheese, grains, mold on foods, dried bananas, dried milk, rotting potatoes, caramel, fermenting substances, nuts, and mushrooms.

Signs of infestation include the presence of buff-colored, pinkish, or grayish dust around food, an unpleasant or minty odor, or—if you can see them—the animals themselves.

Cheese Skippers

Also known as ham skippers, these oddball acrobats are the maggots of a filth fly. They tend to infest cheese, cured meats, and fish. The term *skipper* refers to the fact that by fastening its mouth hook on its abdomen, and then suddenly releasing it, a cheese skipper can leap as far as ten inches.

Like the young of any filth fly—flies that commonly breed in excrement, garbage, rotten food, and the like—cheese skippers can cause serious digestive illness. Some humans who encounter this pest are even more peculiar than the insect. Thinking that the maggots infest the most delicious parts of the cheese, these individuals eat the rotten food—disease-carrying pests and all.

Adult cheese skippers are small, black or bluish-black in color, with bronze tints on the thorax and face. Their preferred targets include cheese, pork, ham, beef, smoked fish, brine-cured fish, and animal carcasses. Signs of infestation include maggots and/or flies.

THE HAZARDS OF PANTRY PESTS

These tiny animals can be much more than a nuisance. Their cast-off hairs can be irritating to the mouth, throat, and stomach, and can cause serious digestive problems. They may also carry highly carcinogenic compounds, especially if the food they're feeding on is stored in warm humid conditions.

HOW PANTRY PESTS MAKE THEMSELVES AT HOME

Before our ancestors learned to store food, pantry insects lived in the nests of seed-eating birds and rodents, in beehives, under tree bark, in wasps' nests, and around any cache of seeds. They also ate dead vegetable and animal matter. Our increased knowledge of food-keeping hasn't changed their habits; it has just provided these nuisances with nearly perfect living conditions.

Chances are that you brought your freeloaders into your kitchen with the groceries, maybe in damaged packages of breakfast cereal or pet food. Or perhaps you didn't notice that the box of raisins you put in your shopping cart had tiny scrape marks or nearly microscopic holes. One little-known fact is that our government agencies allow a few insect eggs and parts to be present in just about any food product we buy. It is impossible for food processors to keep every last one of these critters out of their products. As a consequence, a few eggs may hatch and develop in foods we keep for a long, long time. Sometimes we move cereal or flour or pet food to the back of the shelf and forget to use it again—until it's too late.

Other harborages of these pantry creeps are old furniture, draperies, bedding, rugs—any items made with animal fibers.

Bird, wasp, and rodent nests, as well as nearby dumps and the byproducts of food-processing plants, are also possible sources of these pests. They have many routes for infiltrating a home from the outside—under doors, through and around torn or loose screens, down fireplace chimneys, and around utility pipes.

Weeks, maybe months, after the earliest arrivals sneaked in, their numbers have multiplied to the point where you know you have a problem. Food shelves—generally dark, possibly humid, not too cold, free of predators, and holding a huge stock (by insect standards) of food—provide an ideal environment for these animals. According to Carol Sutherland, if you suspect that your cupboards are hosting some of these undesirables, you can quickly check by running your finger along the inside edge of a flour canister. "If you pick up something that looks like little dust bunnies," she says, "you've probably got an infestation." These are probably the remains of stringy webbing or insect cocoons, sheltering pupae soon to emerge as adults. Unfortunately, by the time you notice this, the insects may have spread to other packages on your shelves, and the often slow, expensive process of getting rid of them has to begin.

Pantry insects like to nest near the underside rims of cans, so to be thorough, check these. You should also keep an eye on any dried flower or fruit arrangements you have, because these insects are attracted to them and can readily find their way from your decorations to your stored foods. Watch stores of hobby items such as feathers or pieces of leather because these can harbor insects if left unused for a long time.Whatever these creatures are eating, getting rid of them can be a headache.

If a container of flour or cereal has lumps of food clinging to the sides, you probably have flour beetles. Or you may find food that is crisscrossed with webbing. By the time such signs appear, insect young have spun their cocoons and are resting before emerging as egg-laying adults. Or the food may be dirty with droppings, cast-off larval skin, and partly chewed food particles. In heavy infestations, some insects give off a peculiar odor. For example, flour beetles have a distinctive, often foul scent, while the smell of crushed mites may remind you of mint.

You don't need to wait until you see crawlers in the corn flakes to know that there is trouble in the kitchen. Early warning signs can alert you to a potential or recent invasion. If you find packages with any sign of damage on the supermarket shelf, leave them there. And let the store manager know that the market's warehouse inspection procedures need to be tightened; you'll be doing other customers a big favor.

SHUT THEM OUT

Halting pantry pests before they invade, settle in, and raise their families in your food is a lot easier than trying to evict them. Here are a few simple routines that will help keep insects and mites out of your cupboard.

Outdoors

❑ Eliminate any bird, wasp, or rodent nests near your home. These often harbor pests that prey on stored food. If a pair of birds starts to carry twigs or lint to a vine growing on your wall or a tree overhanging the roof, chase them off. Watching a family of nestlings develop is fascinating, but it's no fun to cope with the seed-eating pests that can migrate from their nest to your kitchen.

❑ Eliminate white flowers in your garden, especially crape myrtle blossoms. Adult carpet beetles seem to be drawn to them.

❑ Check all cut flowers for signs of insects before bringing them inside.

Indoors

❑ Before putting any packaged foods into your supermarket cart, inspect them carefully. If package flaps are loose or the cardboard is punctured, no matter how tiny the holes—*especially* if the holes are tiny!—don't take the item. Be especially fussy about all cereals, pasta, pastry, cake, biscuit mixes, dried fruits, legumes, powdered milk, bird seed, and pet kibble. If you find a damaged package when you get the groceries home, take it back immediately, before any possible infestation has a chance to spread.

❏ Buy legumes in moderate amounts, and rotate your stocks.

❏ Keep your kitchen cool and dry. If possible, ventilate your cupboards. Many stored-food pests need high humidity and temperatures between 75°F and 85°F.

❏ Place packets of silica gel on pantry shelves to help hold down humidity. Silica gel, which is nontoxic, can be bought at hardware and builders' supply stores. Set the granules in a jar covered with a punctured lid. When the granules turn pink, they have absorbed all the moisture they can. You can reuse the granules by drying them out in the oven until they turn grayish-white again.

❏ Make sure your kitchen walls, ceiling, and floor are free of cracks and holes where insects can hide and lay their eggs.

❏ Keep your kitchen drawers and cupboards clean and dry. Some authorities claim that scrubbing shelves with soap and water simply shifts insect eggs to more inaccessible cracks, where they become ready to hatch as soon as the cabinets are dry. Vacuuming the insides of your cupboards at regular intervals to pick up crumbs, or dusting them with a soft brush, gives more effective control. Don't overlook the undersides of shelves, where pest young often hang their cocoons.

❏ Wash flour, sugar, and other canisters thoroughly—inside and out—before refilling them.

❏ Rotate your food stocks, using up older ones first. Do not buy more grain foods than you expect to consume in a short time, especially in the summer. If you don't bake or eat hot breakfasts in the summer, use up any cereals that need cooking and any baking supplies before the hot weather sets in. It is probably safe, though, to keep sound, unopened boxes of cereals on the shelf in warm weather, because in most cases, cartons have been treated with a repellent approved by the U.S. Food and Drug Administration (FDA).

❏ Don't mix old foods with new foods. If you do, you could spread an infestation from one to the other.

❏ Check your shelves from time to time for opened, forgotten packages of food products. These are often the major harborage for these pests. Cereals with only one insect in the package may be swarming with them a month later. In a well-run food warehouse, inspection of stock is ongoing, and at the first sign of wildlife, the food is destroyed or treated. Since the reproductive cycle of some insects can be less than a week when temperatures are high, check your staples at least once a week in summer, and once a month in winter. While looking over each item, shake or stir it vigorously to disrupt any reproductive process that may be going on.

❏ Store all nonrefrigerated foods in tightly closed containers. Be especially careful about pet food, a favorite of many pantry pests. A tightly lidded garbage can locks them out of kibble; for smaller amounts, a tin cracker box or airtight plastic container works just fine. Keep spices and herbs in screw-top jars or reclosable tins. Since all of these pests (with the exception of adult moths) have sharp mouth parts called mandibles that can penetrate hard protective shells, plastic bags are useless for long-term food storage.

❏ Shake out the contents of one spice container at a time onto a sheet of white paper, and watch for any movement. A magnifying glass can help you locate insect parts.

❏ To protect against cheese skippers, wrap cheese in cheesecloth and dip it in melted paraffin. Cover and refrigerate all meat and cheese.

❏ Be careful when buying used furniture, bedding, or draperies. The animal or vegetable fibers these often contain may be sheltering pantry pests.

❏ Make sure all your windows have tight-fitting screens.

❏ In times past, careful homemakers routinely dropped a bay leaf or two into their stocks of grains to repel insects. Actually, this may keep out a stray bug or two, but can't be relied on to suppress insects that come into your home with the groceries. Some of these pests, like the eat-everything cigarette beetle, may like bay leaves—or cinnamon, cloves, mint, and any other reputed repellent. The homemakers who found herbals reliable probably kept their pantries immaculate. However, if it makes you feel more secure to have bay leaves in your cereal, put them in a cloth bag. Otherwise, the brittle leaves will crumble into the food. And who wants to eat bay-flavored oatmeal?

WIPE THEM OUT

Pests of stored foods are among the most annoying and stubborn of insect nuisances. Because an infestation of these pests can cause serious illness, it should not be taken lightly. Don't shrug pantry pests off as just "supplementary protein," as one well-meaning friend suggested to me. Once they are feeding on your staples, you will need persistence, patience, and even some ingenuity to clear them out. The approaches described below may take some time but, if you follow them carefully, you can get rid of your tiny boarders.

❑ Persistence and watchfulness. As soon as you notice any sign of these insects, look over all the foods in your cabinets, including any unopened cardboard boxes and cellophane or plastic bags. If you find even one insect, webbing, larvae, or tiny punctures or scrapings on the outside of a container, put the food in a plastic bag, and close it securely with a knot or rubber band. Put the bag in a garbage container outside—preferably in hot sun, so that the heat will kill any organisms that may be present.

❑ Thorough cleaning. Empty your pantry completely, and vacuum all surfaces thoroughly. Use your crevice tool to suck up bits of food that have fallen behind shelves and into cracks.

❑ Quarantine. Place any suspect foods—those that were near the site of the suspected infestation, whether or not there are any signs of trouble—in clear plastic bags for a few days so you can watch them for signs of insect activity. (If you have a garage or some other protected ouside area, that would be a good holding spot.) If life seems to be developing in any of them, throw them out as indicated above. Quarantine food in an unopened package by putting the whole package into a plastic bag. Any insects emerging from the container will thus be visible.

❑ Cold. If your refrigerator is large enough, store the rest of your opened food stocks in it for a few weeks. Most species die after prolonged exposure to 40°F or below. (An exception is the spider beetle, which can withstand great cold.) If you have a deep-freezer with a steady temperature of 0° to 5°F, you have an all-purpose pesticide. Two to three days in deep cold will kill most pest species, and three weeks will destroy all their life stages.

❑ Heat. If you don't have a freezer, spread your staples in thin layers on flat pans with raised edges. Heat them for two hours in your oven at about 125°F. Don't let the heat rise much over that, or you will destroy some nutrients. If you have a gas oven, just turning the pilot light up a bit may do the trick. Use an oven thermometer to check the temperature from time to time. If it rises above 130°F, prop the oven door open a few inches to bring it down.

❑ Boiling water. If you find wildlife in raisins, prunes, apricots, or other dried fruits, and you want to salvage these expensive foods, drop them into boiling water for one minute. Rinse them off and dry them before putting them in a tightly closed container.

❑ Carbon dioxide (CO_2). Use dry ice, which is solidified carbon dioxide. It is usually available at ice-making plants. Some ice cream stores and supermarkets also sell it. Be careful when working with dry ice. Do not handle it with your bare hands (it can give you a bad freezer burn), but wear thick gloves or use tongs. Put a layer of the food you are treating in a clear plastic container. Place a cube of dry

A Grain-Processing Company
That Uses No Chemical Pesticides

In the Texas panhandle, not far from Amarillo, Arrowhead Mills annually processes tons of grains and legumes, much of them stored in outside bins all year round, without using pesticides. How do they do it? According to the company's president, Boyd Foster, with scrupulous cleanliness, cold temperatures, and mechanical jolting.

Before each new batch of grain is poured into the machinery, all milling equipment is thoroughly cleaned and vacuumed. Storage warehouses are kept at a steady 40°F. The company also has an ingenious solution to the problem of stored-food pests in its outdoor bins, where unprocessed foods are kept for months. Fans at the bottom of the bins run all through the frigid panhandle winter, pulling icy air through the mass of grains and legumes. So deep is the chilling that the food remains protected for the rest of the year.

Harvested grains and beans often carry insect eggs that don't hatch until the food is in milling or storage, or even until it reaches your kitchen. To destroy such eggs, Arrowhead Mills has raised the stirring and shaking of stored foods to a fine art. With a device called an *entoleter*, they spin the stocks at a dizzying speed, ending with a sharp jolt. Any insect eggs are completely crushed.

ice on the food, and then fill the container with more of the food. A ½-inch cube is enough to treat a pint jar, a 1-inch cube can clear out a quart jar, and a 2-inch cube can take care of a two-quart container. However, CO_2 kills only adults and larvae; eggs and pupae are unaffected by it.

After several hours, when the dry ice has evaporated, seal the container completely. Treating foods with carbon dioxide does not change their quality. Cereals and grains do not clump. It may take as long as fourteen days to completely kill all life forms.

Several important notes of caution: If the container you are using has a screw-top lid, immediately after adding the dry ice, turn the lid until it begins to tighten, then turn it back to loosen it. *Lids must be loose to allow gas to escape and prevent an explosion.* In a tightly closed container, dry ice can build up a pressure of 200 to 250 pounds per square inch, enough to burst a glass jar, which is why you should never use a glass jar for this procedure. Also, if you use dry ice, be sure to work in a well-ventilated area. CO_2 readily replaces oxygen in the air and can lead to asphyxiation.

If you are dealing with a flour-beetle problem, but only eggs (not larvae or adult animals) are present in a large quantity of flour that you want to save, sift the flour through bolting cloth. This is fine-meshed silk or nylon used by grain processors for the final milling. It is also used by artists and photographers, and can be bought in art and photography supply stores.

Pantry pests can drive a homemaker frantic. Just when you think they're finally gone, they turn up in food you're positive was free of them just a short time before. Food is expensive, and throwing it out is painful. You may be desperate enough to consider spraying your pantry with a chemical. Don't! No insecticide is safe for such a use. And remember, some of these animals actually thrive on poison.

Your best protection against these mini-pirates is prevention—watchful food buying; keeping your home clean, cool, and dry; buying food in moderate amounts; and keeping a close eye on your pantry and its contents.

If some insects manage to slip through your defenses, the safest way to get rid of them is to:

❑ Throw out all infested items.

❑ Quarantine all susceptible foods, even if they show no signs of insect activity.

❑ Heat or freeze all susceptible foods.

❑ Buy new stocks in the smallest amounts possible for a few months, and use them up quickly. Keep them completely separate from foods already in your kitchen.

Sticky traps laced with pheromones are useful for trapping insects remaining after the infestation source has been removed. Traps for the Indianmeal moth only can be bought in retail stores. They will not attract the female meal moth, but they will reduce her egg-laying ability if they catch males before they have a chance to mate. These traps will not attract beetles, however.

You have probably noticed advertisements for so-called electronic or ultrasonic pest repellers that are promised to keep six-legged invaders out of your house forever. Some years back, these deceptive ads promised to repel insects, but after a series of tests by the U.S. Environmental Protection Agency (EPA) showing them to be completely ineffective, the ads were withdrawn—sometimes after legal action. They seem to be appearing again, only now they don't mention "insects"— just "pests," thus skirting government sanctions. If these gadgets really worked, managers of food-processing plants, storage warehouses, and restaurants would buy them by the millions. Controlling food pests is an ongoing headache in all food-handling facilities. Only by using relentless sanitation can the pest population be held within tolerable limits. "Sanitation," says one pest control authority, "is pest control."

When you first find these pests in your foods, they will seem to have dropped from the blue. No telltale buzz or whine announces them; and by the time they catch your attention, they are well established, well fed, and rapidly reproducing.

Of course, you would like to get rid of them right away, but chances are you can't. Most likely, they sneaked in weeks before you noticed them, and it may take weeks before you are free of them. Have patience. By following the suggestions detailed in this chapter, you will eventually clear them out safely and, alert to the early warning signs, will be able to avoid future infestations.

5. Rats and Mice— There's No Pied Piper

On a cold day in December 1818, Pastor Joseph Mohr sat down at the church organ in Oberndorf, Austria, to practice for the Christmas Eve Mass. He pressed the keys and pumped the pedals, but the sturdy old instrument gave out more rheumatic wheezes and sighs than musical tones. Mice had chewed the bellows full of holes. Having no time to repair the organ, and determined not to have the Mass without music, he asked his friend Franz Gruber to compose a simple melody for guitar while he wrote the lyrics. Borrowing an *Andante for Winds* from Mozart, they worked through the night on December 23, and on Christmas Eve, the two friends gave the first public performance of *Silent Night, Holy Night.*

Except for a distant relative named Mickey who has made it big in Hollywood, that was probably the only worthwhile thing that a domestic rodent has ever done for humanity.

Mice—and their larger kin, rats—have plagued us for thousands of years. They have invaded our dwellings, food stores and warehouses, restaurants, museums, offices, and transport vehicles—even our books. So intimately are these animals bound up in our lives, that they are called "domestic" rodents. If you object to mice and rats sharing billing with your dog or cat, you could use the term *commensal*, which means something or someone that shares your table—a notion you may like even less.

TYPES OF RATS AND MICE

Among the most successful mammals, mice are the most widely distributed of this major group. Both rats and mice probably originated in Asia, but they have been carried in ships, caravans, trains, trucks, and even airplanes to the four corners of the globe.

In the United States, the roof rat is the one most likely pattering over wires and fences along the West and Gulf coasts, and in the southeastern states as far as

North Carolina. The Norway rat (also known as the sewer, wharf, or brown rat) and the house mouse are the most common rodent pests throughout the rest of the country. The larger, heavier Norway rat is the one generally found in human structures; the lighter, more agile roof rat can prosper independently of human habitation. Figure 5.1 illustrates the differences in appearances of the three species.

Albino rats, which are used in many laboratories and often kept as pets, are Norway rats. When they escape or are intentionally released, they quickly establish themselves as community pests, and are as destructive and dangerous as their darker cousins.

It has been said that in densely populated southern California, there is one rat for every human being, many of them in well-kept neighborhoods. "Is that *all?*" commented one official with the Pasadena Health Department. And, indeed, this statistic may be questionable; the real number may be much higher.

Norway Rats and Roof Rats

Both Norway rats and roof rats find human habitation very comfortable. With its tendency to burrow, the Norway rat is ideally suited to a rural environment. The roof rat, an agile climber, is more at home in the city, with its wires and tall buildings.

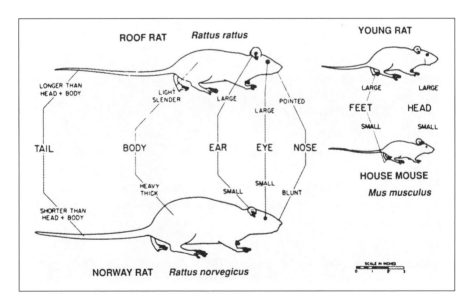

Figure 5.1 The pictures above identify some of the most common types of domestic rodents. The Norway rat is the most common in temperate zones; the roof rat in the tropics and subtropics. The house mouse thrives everywhere.

Source: Control of Domestic Rats and Mice (Atlanta: Centers for Disease Control, Homestudy Course 3013-G, Manual 11, 1982).

Charactistics of Norway rats include the following:

• They burrow under buildings and piles of lumber to travel to and from their nests. The burrows are up to 15 inches deep and $2\frac{1}{2}$ to 4 inches wide, and may be as long as a city block. Escape holes called bolt holes, set under grass or boards, provide rodents with quick escape routes.

• They nest in damp places such as stream banks, garbage dumps, and marshes.

• They eat both vegetables and meat, and are smart enough to prefer fresh food to spoiled.

• They are nocturnal.

• In their nine-to-twelve-month lifespan, females can produce four to seven litters of eight to twelve young each.

Charactistics of roof rats include the following:

• They nest in woodpiles, pyracantha (firethorn) bushes, thick growths of ivy, and palm trees and other evergreens, all of which are commonly found in the suburbs.

• They eat insects, berries, grains, nuts, fruits, vegetables, and manure. They need less water than Norway rats.

• They are nocturnal.

• In their nine-to-twelve-month lifespan, females can produce up to six litters, with six to eight young in each.

Both Norway and roof rats are strong swimmers, able to paddle around for hours, and are often found in sewer systems. Even in affluent neighborhoods, they can swim through floor drains and toilet seals. In many large cities, rats use older sewage systems for highways. Where storm and sanitary sewers are combined, the sewer rat problem is apt to be worse, especially if garbage disposals are in general use and food scraps mingle with waste water.

The House Mouse

Because mice originated in the dry grasslands of Asia, they are well adapted to life without a steady water supply, and can survive long periods in closed boxes, trunks, and barrels, where they have plenty of time to wreak havoc. Like rats, however, mice are very good swimmers.

The house mouse is the species most likely to infest your home. Their characteristics include the following:

• They can live comfortably in a sack of grain or other dry food without visible moisture.

- They do burrow, but their tiny trails are nearly impossible to find.

- In their lifetime, females may produce seven or eight litters, with five or six young each.

Rats rove up to 1,500 feet from their nests, whereas mice wander just a few feet from theirs. Between nests and food sources, both establish harborages (resting places where they can hide and eat). Neither type of animal sees well, so they tend to run along walls, where their long guard hairs make contact and guide them.

THE HAZARDS OF RODENTS

Throughout history, domestic rodents have tormented us. The bubonic plague (the dreaded Black Death) is transmitted by fleas that infest rodents. Plague has killed millions of people since at least the sixth century A.D. The Pied Piper was said to have charmed away the rats in the town of Hamelin in the late 1200s, perhaps in the beginning of the Great Plague, which nearly depopulated Europe in the following century. Despite effective modern antibiotics, thousands of people around the world still die of this disease every year, though it is not a major problem in the United States.

In this country, rats and mice are more apt to transmit typhus, dysentery, infectious jaundice, food poisoning, murine (fleaborne) typhus, trichinosis, rickettsial pox (a disease similar to chickenpox), and rat-bite fever, which, despite its name, is also transmitted by mice. Dale Bottrell, author of *Integrated Pest Management*, says that an estimated 60,000 Americans, mostly infants in their cribs and bedridden elderly, are bitten by rats every year. More vicious than mice, rats will attack if cornered, and can chew off fingers and toes. Once a rat has bitten a person, it will probably do so again. Mice also carry pathogens for a form of meningitis; up to 60 perecent of mice may be infected and an estimated 9 percent of humans carry antibodies that indicate exposure. House mice do not carry the deadly hantavirus; however, deer mice and white-footed mice, which are increasingly found in suburban areas, do. Health authorities estimate that in the past hundred years, 10 million people have died from rodent-transmitted diseases.

ECONOMIC DAMAGE CAUSED BY RATS AND MICE

The economic harm caused by rats and mice is almost beyond estimation. Rodents destroy as much as 20 percent of the world's food crops every year. In India alone, the grain they consume yearly could fill a train 3,000 miles long. In the United States, the damage caused by rats and mice is estimated in the hundreds of millions of dollars per year.

Rats gnaw constantly to keep their incisors, which grow about an inch a year, at a manageable length for feeding. They gnaw into upholstery and other fabrics, and chew books and paper to get nesting materials. If they file their teeth on electrical wires or matches, they can set homes on fire. Rats have put out the lights at a London airport, burned down warehouses, flooded computers, and electrocuted cattle. Exerting an incredible 24,000 pounds per square inch with their incisors, they can gouge holes in gas pipes, asphyxiating whole families.

Mice are nearly as destructive. Besides ruining crops and food stocks, they can nibble family records, valuable paintings, and manuscripts. Because mice urinate constantly, they contaminate everything they touch. They can make themselves at home in electrical appliances, causing dangerous short circuits.

HOW RATS AND MICE MAKE THEMSELVES AT HOME

Many city-dwellers live in apartments all their lives without ever meeting up with a rat or a mouse. Although blighted urban neighborhoods carry large rodent populations, well-kept buildings are usually free of such wildlife.

On first moving to the suburbs, city people are often delighted with their new neighbors—deer, rabbits, cedar waxwings, and racoons. Their delight may quickly fade, however, when they find that snakes, skunks, rats, and mice also live in the vicinity.

Rodent infestation of suburban housing follows a marked pattern. Many housing developments are built on former orchards (or, in the Sunbelt, citrus groves), where rodents may be plentiful. For a few years, the new homes are rodent-free, but then pools are put in; bird feeders are erected; shrubs, trees, and vines thrive; and the stacks of items stored in the attached garages, basements, or attics grow. Shake roofs develop holes, vent screens deteriorate, and the house settles, opening small cracks under the eaves, in the foundation, and around doors and windows.

Within ten to twelve years, the roof rat population explodes, and as many as 20 to 25 percent of homes in suburban areas may be infested. Twenty to forty years later, with trees and shrubs mature and some of the homes poorly kept, the whole area is open to vermin infestation.

The most vulnerable homes are those that are older and neglected, and therefore have many points of entry and hiding places for small animals. Poor disposal of food waste aggravates the problem, as do nut trees, neglected fruit trees, outbuildings, woodpiles, and thick vegetation.

Severe weather conditions can make the problem much worse. In 2002, California and other states of the Southwest were suffering one of the severest droughts in history. Thirsty rodents, desperate for water, came out of their hiding places and invaded homes and businesses in wealthy communities like Beverly

Hills and Santa Monica. Several restaurants in Santa Monica had to close until the pests could be driven out. These were the ones that made the headlines. Who knows how many other places have had the same problem?

SIGNS OF RODENTS

"Out of sight, out of mind" describes most people's attitudes toward domestic rodents. If we don't see them, we think they aren't there. But there are clues to their presence that should make you take decisive action. Either one or several of the following signs can mean that you have some undesirable four-footed boarders:

• Sounds. At night, you may hear the pattering of paws, climbing sounds in the walls, squeaks, or churring sounds. Mice give a little whistle, rather like a canary. To be positive that these are rodent noises, after you have been out for an hour or two in the evening, enter your house very quietly and stand still for a little while. If rodents are in the walls, their sounds will gradually resume.

• Droppings. By the time you hear rodent noises, you may have noticed droppings in your kitchen cupboard or on the counter top. They will be heaviest along rat or mouse runways and near your food supplies. Variations in the size of droppings can indicate an established colony, with older and younger animals.

• Urine. Some authorities recommend flashing black light in the area to detect rodent urine, which fluoresces under ultraviolet light. So do other substances, however, so this is not an infallible clue. A strong flashlight with a red filter is not apt to send a rat scurrying out of sight. In long-term infestations, mice leave a "urine pillar," a pile of urine, grease, and dirt.

• Gnawings. Freshly gnawed wood, lighter in color than the wood around it and with small chips of the same color scattered about, are a strong rodent sign. Since these animals will smooth over a hole through which they have to pass, scratchy, sharp gnawings indicate recent activity.

• Grease marks. Greasy stains from rodents' fur along runways and wherever they swing past the angles of joists and ceiling rafters are another telltale clue. Such marks can also rim gnawed holes, follow along pipes and beams, and soil the edges of stairs. Mouse smudges are much smaller and harder to see than those of rats. Norway rats run along the floor, so smudges higher up indicate roof rats.

• Odors. In large numbers, rats and mice give off a peculiar pungent smell. If your house has a persistent odor that you cannot identify, you would be wise to look for rodents.

• Excited pets. If your cat or dog repeatedly sniffs and paws at the floor in front of one wall spot, near a kitchen cabinet, or among stored items in the garage or basement, you should, as the saying goes, "smell a rat." This is especially true if the invasion is recent.

• Actual sightings. If you see a live rodent, you'll know you have a problem. Rats are secretive and nocturnal, so if you spot one in the daytime, you can be sure that there is a large population of them on the premises and in the burrows and nests, forcing that individual out. For every rat you see at night, you can assume that there are ten or more in the general area. Mice are active in the daytime, so seeing one in the kitchen in the morning may or may not mean that you have a colony in the house.

• Food. If you think you are harboring rats or mice, putting a few pieces of dry dog chow out overnight in an attic, basement, or room forbidden to your pets may confirm your suspicions. If the food is gone the next morning, you should look for other signs.

• Dead rodent remains. If you find a dead rat or mouse in or near your home (assuming you don't have a cat that likes to bring you "presents"), notify public health authorities at once. Ask if a poisoning program is underway. If the answer is no, the animal may be diseased. *Do not* handle it with your bare hands. Use tongs, rubber gloves, or two sticks to put it into a plastic bag. Tie a knot at the top of the bag, and throw it into the garbage or bury it. (For further details on the disposal of rodents, see page 90.)

• Tracks. Outdoors, rat runways are clean packed-earth paths two to three inches wide, with tracks of claw marks in the dust. They are readily seen along walls, under boards, and behind stored objects and litter. You may also notice paw marks in the mud around a puddle. It's easier to see tracks at night with a flashlight held sideways so that the indentations cast a shadow. (See Figure 5.2.)

If you suspect that there is a rodent runway outdoors, but the marks are not clear, set a pipe or long, narrow box with its ends cut off along the path and dust the inside of it with flour or talc. The powder will be protected from the elements, and any rodent running across it will announce itself.

To detect rodent paths indoors, sprinkle flour or talc wherever you think the animals are traveling. If you you have spotted a runway, you will soon see giveaway tracks.

Both rats and mice can be persistent problems, but mice are even more adaptable to life with human beings than rats are. They have been found flourishing in the upper stories of high-rise buildings even before the structures were finished, probably living on construction workers' lunch scraps. These tiny mammals are highly intelligent. In an experiment conducted in 1968, one species of mouse learned to operate twenty-two separate switches that controlled the lighting in their cages, temperature, and food and water supplies, as well as operating the

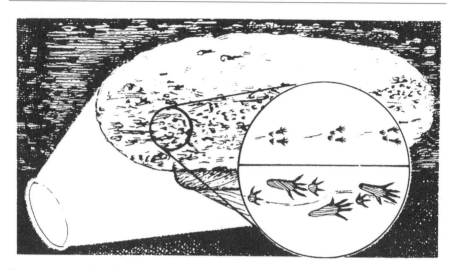

Figure 5.2 Tracks of Mice and Rats

Source: Control of Domestic Rats and Mice (Atlanta: Centers for Disease Control, Homestudy Course 3013-G, Manual 11, 1982).

running cages—all this in only three days. You have to be alert and persistent to outwit this clever little intruder.

With rodents, as with all pests, shutting them out is always the first priority. However, if you fail to get rid of rodents before tightening up your house, and your housekeeping, they are likely simply to find or create new ways to get back in. Once you have cut off rodent food supplies, destroyed their harborages, and killed those that have set up housekeeping with you, you can concentrate on barricading your home against them.

Pest control professionals call this process *rodent-proofing*. To make your home inaccessible to these pests, you need to know what rodents can do. These are fantastic acrobats.

Be aware that both rats and mice can:

• Go through very small openings. A rat needs slightly more than a one-half inch of space; a mouse, only one-quarter inch to gain entry.

• Walk along thin horizontal wires or ropes and climb vertical ones. A mouse can travel upside-down on mesh hardware cloth.

• Climb along the outside of a vertical pipe three inches in diameter.

• Climb along the *inside* of a vertical pipe one and one-half to four inches in diameter.

• Climb *any* pipe or conduit within three inches of a wall.

- Climb up a smooth surface if there is a pipe nearby.

- Climb a stucco or brick wall.

- Crawl along any horizontal pipe or conduit.

- Drop as much as fifty feet without being killed. Mice have fallen from a fourth floor without injury. Both mice and rats have an excellent sense of balance.

- Travel on extremely narrow ledges.

- Gnaw through wood, paperboard, cloth sacks, lead pipes, cinder blocks, asbestos, aluminum, adobe brick, and concrete (before it's fully hardened).

Rats can:

- Leap up three feet from any flat surface.

- Jump four feet across a flat surface.

- Jump horizontally at least eight feet out from a fifteen-foot height.

- Reach out about thirteen inches without jumping.

Both rats and mice are strong swimmers—even roof rats, which prefer drier environments. All have poor vision, but all of their other senses are keen.

How can one keep such phenomenal athletes out of a building? By closing off all possible entries and routes of access.

SHUT THEM OUT

Rodent-Proofing New Construction

If you are remodeling your home or building a new one, there are a number of measures you can take to prevent mice and rats from gaining access after the project is finished. These include the following:

❏ Have your contractor nail galvanized sheet metal cut to fit between studs along the sill. Or fill these areas with a good grade of rich cement. (See Figure 5.3.)

❏ Build a curtain wall. Since rats burrow perpendicularly to walls and do not try to go around them, a curtain wall that extends downward and has a horizontal lip stretching out about twelve inches from the base prevents both rat infestations inside, and underground burrowing that can cause walls to collapse.

❏ For a home on pilings, a curtain wall of one-quarter-inch wire mesh will bar rodents. (See Figure 5.4.) Another type of rodent stop, illustrated in Figure 5.5, is suitable for a house on a concrete foundation.

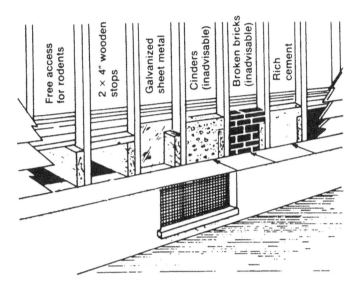

Figure 5.3
Methods of
Excluding Rats
from Double
Walls

Source: Rodent-Borne Disease Control Through Rodent Stoppage (Atlanta: Centers for Disease Control, 1976).

❏ Choose rodent-proof materials. Use steel and cement construction wherever possible, with a concrete foundation and basement floor. Be careful to avoid partially excavated cellars, and any other unexcavated areas under the building and deck. Drains and ventilators should be protected with heavy-gauge screening. Fireplace flues should fit snugly and be kept closed when not in use.

❏ Make doors and windows tight. Make sure all doors and windows fit snugly and are flashed with metal; screen doors should close automatically within three or

Figure 5.4
Wire Mesh Rat-Stopper for a
House Without a Foundation

To prevent rats from burrowing underground, use one-quarter-inch mesh wire screening around the bottom part of the house, tightly fastened to the siding and buried twelve inches deep and twelve inches out from base of building.

Source: How to Control Rats on Your Property (Pasadena, California, Health Department leaflet, undated).

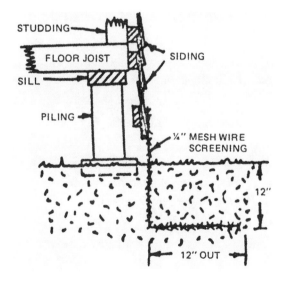

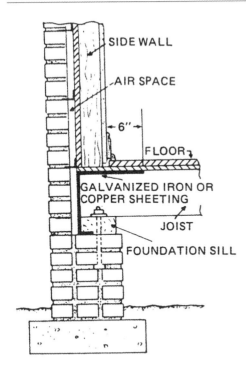

Figure 5.5

One Type of Recommended Rat-Proofing Between Floor and Foundation in Residential Construction

Source: Harold G. Scott and Margery Borom, *Rodent-Borne Disease Control Through Rodent Stoppage* (Atlanta: Centers for Disease Control, 1976).

four seconds. Hardware cloth covering basement windows should be of one-quarter-inch mesh.

Rodent-Proofing an Existing House

If you are living in the house and have rodents, inspect your property carefully, starting outside.

❏ Check your yard for rat burrows. A well-meaning neighbor may suggest that you gas any that you find. *Don't.* This is a job for a professional pest control operator. The gases used are lethal and, since burrows often extend under a house, they can poison human occupants. There are, however, two safe steps you can take to destroy rodents in their burrows:

1. Plug all holes, including bolt holes, with cement. Mix the fresh cement with broken glass to keep rats from gnawing through it before it sets.

2. If your soil is a tight clay that holds water, flood the burrows. This either drowns the rodents or drives them out so they can be clubbed to death.

❏ Barricade your roof. Getting to your roof is no problem for rats, whose runways have been traced as high as eleven stories. Remove any vines growing against the building, and prune away all overhanging tree limbs to at least eight feet from the roof. All roof vents, whether for kitchen fans or plumbing, and chimneys should be rat-proofed with metal screening, as shown in Figure 5.6.

❑ Guard pipes and wires. All pipes and wires coming into your house should be fitted with metal guards. (See Figure 5.7.) Nails should be set flush with the surface of the barrier so that no rodent can get a toehold on them. They should be far enough apart so as not to serve as a rat ladder (remember, a rat has a thirteen-inch reach).

❑ Repair cracks and holes. See to it that every crack and hole along the foundation or under the roof is completely patched with masonry. Where walls have settled, or if you live in an earthquake-prone area such as the far western United States, watch for cracks in ornamental brick—ideal entryways for rodents. Seal all openings around pipe and wire entries or cover them with sheet metal cut to fit, as shown in Figure 5.8. If you use cement, mix it with broken glass. Coarse steel wool can serve as a temporary closure, but since it soon rusts, you shouldn't depend on it for very long.

❑ Protect basement windows. Basement windows should be protected with 17-gauge ½-inch hardware cloth set in removable frames. To be doubly sure, fit 16-by-20-mesh fly screening behind the hardware cloth.

❑ Cover attic vents. Cover attic vents firmly with ¼-inch mesh hardware cloth.

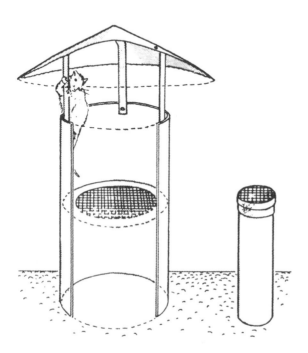

Figure 5.6
Stoppage of Vent Pipes

Source: Harold G. Scott and Margery Borom, *Rodent-Borne Disease Control Through Rodent Stoppage* (Atlanta: Centers for Disease Control and Prevention, 1976).

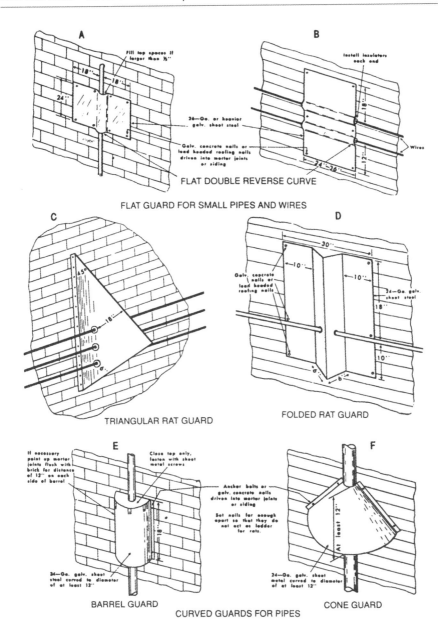

Figure 5.7 Types of Rat Guards for Pipes and Conduits

Thin conduits can be blocked with flat guards for small pipes and wires, triangular rat guards, or folded rat guards (pictured in A, B, C, and D, above). Barrel and cone guards (E and F) are suitable for larger installations.

Source: Harold G. Scott and Margery Borom, *Rodent-Borne Disease Control Through Rodent Stoppage* (Atlanta: Centers for Disease Control, 1976).

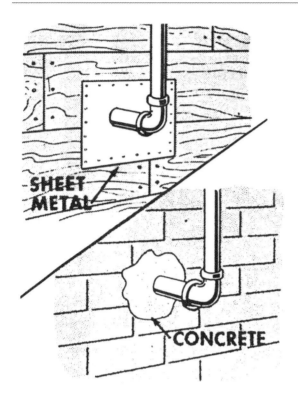

Figure 5.8
Stoppage of Openings
Around Pipes

Source: Harold G. Scott and
Margery Borom, *Rodent-Borne
Disease Control Through Rodent
Stoppage* (Atlanta: Centers for
Disease Control, 1976).

❑ Watch letter drops. A letter drop that is less than eighteen inches off the floor should have an opening smaller than $1/2$ inch. If you have a letter drop with a wider opening, install a spring-closing cover on it.

❑ Screen floor drains. Floor drains in your garage, basement, or pool and pool deck should be sturdily screened.

❑ Protect doors. Protect the edges of wooden outside doors with metal channels and cuffs, as shown in Figure 5.9, to make it impossible for rodents to gnaw through them. All doors to the outside, including basement doors, should have metal flashing.

❑ Check thresholds. While looking over your doors, check their thresholds. If they are worn or broken, replace them or cover them with 24-gauge galvanized metal. The distance between the bottom of the door and the threshold should not be more than $1/2$ inch; $3/8$ inch is better.

❑ Use rodent-proof materials. In modifying your home, wherever you can, use rodent-proof materials such as 24- to 26-gauge sheet metal; expanded metal 28-gauge or heavier with mesh no larger than $1/4$ inch; iron grilles of equal gauge with

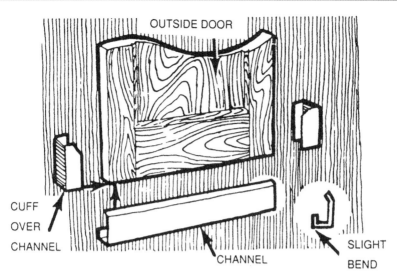

Figure 5.9 Cuff and Channel Rodent Stoppage for Wooden Doors

Source: Control of Domestic Rats and Mice (Atlanta: Centers for Disease Control, Homestudy Course 3013-G, Manual 11, 1982).

slots ¼ or less wide; and perforated metal with ¼-inch holes. Other materials rodents can't penetrate are concrete, brick, mortar, and tile.

The Catch in Rodent-Proofing

In some ways, rodent-proofing can be a Catch-22 situation. When a building has been rat-proofed, its potential for harboring mice increases because fewer animals are competing for food and living space. If your visiting rodents are mice, you will have to be more meticulous about sealing even the smallest openings. Most city building codes, if conscientiously followed, will effectively rat-proof a structure, but they rarely completely mouse-proof it. Thorough sanitation is your best defense against mice. Once their nests and food supplies are eliminated, mice may move away of their own accord.

To keep your home rodent-free, check the structure at least once a year. (See Keeping the Rodents Out on page 82.)

Electronic Repellents

In the past few decades, we have come to think of electronics as a kind of magic wand that makes all things possible. Serious efforts have been made to apply electronic principles to rodent control. For a while, electromagnetic devices were marketed that claimed to drive off rats and mice as well as insects. They soon proved worthless, however, and the U.S. Federal Trade Commission (FTC) has ordered

Keeping the Rodents Out

Constant settling of the ground and drying of the wood can open new entry-ways for mice and rats, while any changes in plumbing lines or electrical wiring can also give these little mammals easy access. Here are a number of checklists to help you keep all possible points of entry plugged.

LOWER FLOORS

❑ Are all openings in the foundation and under siding sealed?

❑ Are all basement windows screened as described on page 78?

❑ Are all floor drains covered with galvanized heavy-gauge strainers?

❑ Are all toilets in good repair?

❑ Are all pipes and wires entering the house protected with rat guards?

❑ Do all screen doors close automatically?

❑ Does the letter drop have a spring-closing cover?

❑ Are all the openings around pipes and wires entering the house sealed?

UPPER FLOORS

❑ Are tree limbs pruned at least eight feet away from the house?

❑ Are attic vents covered with sturdy hardware cloth?

❑ Are all openings and intersections at the roof line sealed?

marketers—and, in some cases, taken legal action against them—to stop their false and misleading claims and to back up any future statements with solid evidence.

Ultrasonic devices seemed to make more sense. Rats and mice have intensely acute hearing. Any loud noise sends them bolting for escape. Environmentalists and food processors once thought such devices held great promise for rodent control. Sad to say, that promise has not been fulfilled. Ultrasound is directional, does not penetrate stored objects, and loses intensity very quickly with distance. And within a matter of weeks the rats and mice get used to the gadgets and return to their accustomed feeding places.

Two verifiable facts behind these devices are that (1) a mouse can be killed by ultrasound generated at extremely high frequencies and decibels, and (2) both types of devices are dangerous to human beings. It is very hard to kill a rat this way. In addition, such high frequencies use a great deal of energy. Tests run by the

Division of Agricultural Sciences of the University of California have all proven negative.(For a further discussion of ultrasound devices, see page 21 in Chapter 1.)

The following, quoted by the Centers for Disease Control in their manual on rodent control, sums up science's evaluation of ultrasound as pest control:

> Ultrasound will not drive rodents from building or areas; will not keep them from their usual food supplies and cannot be generated intensely enough to kill rodents in their colonies. Ultrasound has several disadvantages: it is expensive, it is directional and produces "sound shadows" where rodents are not affected and its intensity is rapidly diminished by air and thus of very limited value.

The technology has also been shown to have no effect on cockroaches.

The rodents themselves have passed the ultimate judgment on ultrasonic repellers. California state health official Minoo Madon once even found a family of mice nesting in one!

Odor Repellents

Certain odors are known to repel rats and mice. Camphor gum and spurge hold off mice, as catnip and dog fennel are said to do. Unfortunately, the effect of these botanicals is only short-lived. They will probably not succeed in places where there are established rodent colonies. Managers of food warehouses and processing plants would be elated if certain odors or electronic devices could drive rodents permanently from their facilities. So far, unfortunately, none has proved to be of much help.

STARVE THEM OUT

At the first signs of rodent infestation, your reaction may be to set out an array of traps or poison baits. Used skillfully and in sufficient quantities, these can quickly reduce the number of rats and mice on your property. However, without rigorous sanitation, in a short time, the survivors' birthrate will soar, reduced competition for food will let more young reach maturity, and you will have as many vermin as before. And since young rats eat considerably more than their elders, the new generation will be even more destructive than the previous one.

Indoors

❑ Your first priority is to tighten sanitation. This includes cleaning out all possible rodent harborages as well as proper handling of stored foods and food waste. By removing harborages and food, you will intensify the survival pressure on the animals. Competition for food will become fierce; dominant rodents will kill weaker ones outright or throw them out of the community so that they become easy meals for predators.

❏ Give them no rest. To reduce available nesting places inside, clean out the debris in your attic, basement, and closets. Piles of old clothes forgotten in a corner, old papers and magazines, boxes kept for no particular reason—all should be set out with the weekly trash. Areas under porches, stairwells, and outside steps frequently collect such rubbish, so be sure to clean these out too.

❏ Store all foods in tightly closed metal, glass, or plastic containers. Cupboards should be kept clean, and floors swept regularly. Wrap leftovers carefully and store them in the refrigerator. A cake-serving storage platter with a lock-on lid protects leftover pastries. Refrigerate ripe fruits and vegetables, and keep ripening fruit in open-air coolers screened with quarter-inch wire mesh.

❏ Wipe up all spills as soon as they occur, especially milk. If your baby falls asleep with a bottle, take the bottle away. Rats like milk, and babies are frequent victims of rat bites.

❏ If you live in an apartment building where one or two tenants are slobs, your own control measures may not be enough. In such a case, alert the management. It's their responsibility to see that the premises are made and kept rodent-free. Fliers and tenant meetings may be all the pressure needed to get things in shape. Building managers should also see that incinerators work properly at all times. If they consume refuse only partially, the half-charred food scraps can support many rats and mice, along with the inevitable cockroaches and pantry pests.

❏ You may be the world's most meticulous householder, but if you live in a community with rat havens such as open dumps and hog-fattening plants, you are going to be coping with rodents. Corral some concerned neighbors, and start to prod City Hall with newspaper and television messages, door-to-door leaflets, and school programs. Don't let up until authorities have replaced the dump with a sanitary landfill and the pig-feeding station has been cleaned up. Don't hesitate to call on your state's environmental protection agency or your county's health department if local authorities are sluggish in this.

Outdoors

❏ Stack building lumber and fireplace logs eighteen inches off the ground so that even if rodents are living among the logs and planks, you will see their droppings and be able to take control measures. Rats hang out in piles of lumber, dense vegetation, stone rubble, dead palm fronds or other plant matter, and ill-kept storage sheds.

❏ Cut any vines way back, and keep shrubs well trimmed. Pull up weeds and dead plants and put them in plastic bags for the sanitation truck. If you have palm trees, have the dead fronds cut down at regular intervals. Get rid of Algerian ivy or any other dense ground cover, a perfect hiding place for rodents.

❏ Put all your kitchen scraps in a plastic bag kept securely closed in a garbage can in the house. At the end of the day, close the bag securely, either by knotting it or by using a rubber band, and dispose of it in a tightly closed large container outside.

❏ Refuse containers should be of twenty- to thirty-gallon capacity, with handles for easy lifting and lids that fit securely. They should be made of heavy-duty plastic or galvanized metal, with recessed bottoms so that no filth accumulates underneath. A 55-gallon drum is not suitable for this purpose, as it will be too heavy to lift when filled, and the lid usually cannot be tightly closed. Store containers on metal or wooden racks or holders eighteen inches above the ground. Racks can be made of steel pipe or bars or weather-resistant wood. Racks and holders like those pictured in Figures 5.10, 5.11, and 5.12 minimize the chance of a roving dog's (or, in the West, a coyote's) knocking cans over, loosening lids, and making the contents available to the local rodent community. Keeping trash containers on a concrete slab is an effective control measure only if you keep the slab clean, avoid spillage, and always keep the lids on tight.

❏ Be sure to pick up any fruit dropped by trees in your yard. Rats and mice feed on fruits and nuts that fall from trees. If you have pets—hamsters, guinea pigs, birds, rabbits, dogs, or cats—store their food in closed metal or heavy plastic containers.

❏ Take measures to eliminate snails, which make good eating for rats. (See Chapter 12 for ways to control snails in your garden, making it less attractive to rats.)

❏ If you feed your dog outside, dispose of any leftovers as soon as the animal is finished eating. Make sure that no pet food is lost under a doghouse, where a rodent can easily reach it by burrowing. Does your dog have a weekly bone or two to keep its teeth healthy? Rodents also like to gnaw on bones; so, when the dog has finished, throw the bones out.

❏ Birdhouses are a frequent lure for rodents, who may eat both the birds' food and their young. Set any bird feeders in an open area; rodents avoid these. Don't put out more seed than the birds will eat in one day, and spread a band of Tanglefoot—a thick, sticky substance—around the pole supporting the birdhouse to stop predaceous rats. And if you enjoy scattering crumbs on the lawn for birds in winter, remember that you may also be helping rodents to survive.

Figure 5.10
A One-Can Stand Made
of Reinforced Steel
Welded Together

*Source: Rodent-Borne Disease
Control through Rodent Stoppage*
(Atlanta, GA: U.S. Centers for
Disease Control, 1976.)

Figure 5.11
A Chained Dogproof Refuse
Container

*Source: Control of Domestic Rats
and Mice* (Atlanta: Centers for
Disease Control, Homestudy
Course 3013-G, Manual 11, 1982).

Figure 5.12
Proper Rack Storage
of Refuse Cans

*Source: Control of Domestic
Rats and Mice* (Atlanta, GA:
Centers for Disease Control,
Homestudy Course 3013-G,
Manual 11, 1982).

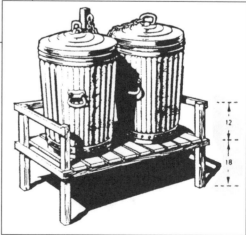

WIPE THEM OUT

Once sanitation in and around your home leaves no food or hiding places for rodents, it's time to eliminate those that are still infesting your property.

Traps—The Best Alternative

Trapping is your safest, surest method—safest for you, and surest to kill the rats or mice. You use no dangerous rodenticides, you can quickly see how successful you have been, and there won't be any dead rodents in places where you cannot dispose of the bodies—in other words, no dead-rodent odors. According to the Bio-Integral Resource Center in Berkeley, California, trapping is best done just before rodent-proofing to prevent rodents being caught inside rat-proofed buildings.

A call to your local health department may bring you experienced trapping help. Some cities and counties can afford enough health inspectors to show you not only how to make your property less attractive to rodents, but also how to do the actual trapping. However, with government budgets tight at all levels, you may have to tackle this job yourself. You could contact a qualified pest control operator for the job. If you are made of sturdy stuff and are not too squeamish, the following discussion will help you choose the right trap for the job and use it in the most effective way possible.

The Snap Trap

One of the most effective devices ever invented for killing rodents is the well-known snap trap, also called the guillotine, spring, and break-back trap. A large California pest control company relies almost exclusively on such traps, monitored by computers, along with rigorous sanitation to keep commercial baking plants and other food processing facilities rodent-free. There are a number of considerations to be aware of in using this trap, among them the size of the trap, the number of traps to use, bait, trap placement, and others.

1. *Choose the right size trap.* If you choose to use a snap trap, be sure to get the right size for the type of rodent you are trying to catch. The spring of a rat trap can completely miss a mouse taking the bait, while a mouse trap won't do much to a rat. Once you have found the right size, choose the model with the strongest spring and the most sensitive trigger.

2. *Calculate the number of traps you need.* When using traps, buy enough of them to mount a quick, decisive campaign. For the average-sized home, twelve or so could do the job. If there seems to be rodent activity in one corner of a particular room, you may have to set five or even ten traps there. For an entire farm, you may need anywhere from fifty to one hundred traps. A reasonable rule of thumb is to use

more traps than you think you have rodents. Using fewer traps for a longer time will simply make the survivors trap-shy. For mice, place the traps no more than ten feet apart since these tiny mammals don't forage very far from their nests.

3. *Use effective bait.* When using traps, your choice of bait is critical. Despite the folklore, cheese doesn't work too well, and pest control operators usually scorn it as a lure. For Norway rats, use a piece of bacon or a slice of hot dog; for roof rats, nut meats, raisins, or a prune. Mice are lured by either a gumdrop or a piece of bacon. Peanut butter smeared on the trigger will attract all species. A piece of cotton, which rodents use for their nests, can also work. Any type of bait can be made more enticing if it is sprinkled with a bit of oatmeal or cornmeal. Tie the bait firmly to the trigger of the trap with a piece of light string or fine wire, or a smart rodent (and some, as we have seen, are very smart) may get the meal without being caught. You may need to try several different kinds of bait before you find one that works. Try using a variety—meat in one trap, cereal in another, peanut butter in a third—until you find one that works well. Baits are most effective if there is little other food around, so keep up your good sanitation routine. Rats are afraid of anything new in their surroundings, so to prevent them from becoming trap-shy, don't actually set a baited trap to spring until the bit of food has been taken for two to three days.

From Berkeley, California—that home of so many original ideas—come two effective baits that have been used successfully by pest control operators there. One placed his traps in students' discarded paper lunch bags. Another baited his traps with frozen pudding laced with sherry. "I got those rats so tight I could catch them with my bare hands," he said.

> Before placing the traps, sprinkle any known runways and harborages with flea powder to kill any parasites that might leave the dead rodents and move on to nearby humans or their pets. Handle the flea powder as carefully as you would any chemical pesticide. (Information on the proper handling of pesticides can be found in Chapter 14.) This is one instance in which a toxic substance poses less risk than the pest itself.

4. *Place traps correctly.* For best results, follow these suggestions:

- Do not place traps where children, pets, or "unknowing feet" may wander.

- Set traps close to the walls where your tracking powder has shown you that the rodents run, behind objects, and in dark corners. Secure them to surfaces so that they cannot be dragged.

- Position traps so that in the natural course of foraging, rodents must pass over the triggers. If a trap is set along a partition, it should extend out from the wall and have its trigger nearly touching the wall. If placed parallel to the wall, use a pair of traps with triggers placed so as to intercept animals coming from either direction. (See Figure 5.13A.)

• To guide the rodents into the traps, set a box near the wall to form a narrow passageway.

• Set the triggers lightly so that the traps will spring easily.

• For mice, place traps no more than ten feet apart since these tiny mammals don't forage very far from their nests.

• When setting traps on rafters, beams, or over overhead pipes, fasten them firmly to these structural parts so that the dying rodent doesn't drag the trap away and die in some inaccessible wall void.

• Nail traps vertically where you see greasy swing and rub marks. (See Figure 5.13.) Do not place traps above food unless the food is covered with a tarpaulin.

• Use multiple-catch live traps for mice, which are available at hardware stores. These can catch several small pests at a time.

• Place three traps set in a row to ensure that a rat trying to jump across them will be caught.

• To avoid injuring children or pets, set spring traps in sturdy bait boxes, and place them where you know the rodents are running. Such boxes, which are available in various sizes, can be bought commercially. Their entrance holes are just large enough to admit a rat or mouse.

• Do not spray pesticides on traps or store them near where these chemicals are kept. Their odors can repel rats if the traps are used later.

5. *Inspect the traps regularly.* This may be a distasteful task, but it is necessary to see how well they are working. Rats are put off by neither the odor of humans nor the smell of other dead rodents.

Don't use petroleum oil to prevent traps from rusting. Its smell repels rats. Use lard or some other animal fat, which will protect the metal and still attract the rodents.

After most of the rats on your property have been eliminated, it may be hard to trap the survivors. (A rat is are especially likely to avoid traps if it has they already sprung one without being injured, so you will need to use camouflage.) Outside, on the ground, sink the trap slightly below ground level and cover its trigger with a small piece of plastic or cloth to keep dirt from clogging the action. Then cover the whole ground area with a layer of fine soil or sawdust. Stones, boxes, or boards can direct an unwary animal along the path to the trap. Indoors, bury a trap in a

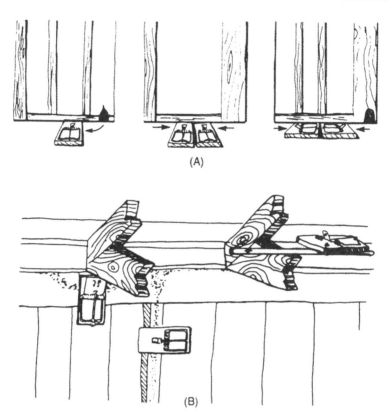

(A)

(B)

Figure 5.13 Effective placement of (A) floor traps: (left) single trap set with trigger next to wall; (center) the double set increases your chances of success; (right) double set placed parallel to wall with triggers to the outside.
(B) overhead traps (particularly useful for roof rats); trap at left modified by fastening piece of cardboard to expand its trigger size. You can nail traps to walls or secur them to pipes with wire or bungee cords.

shallow pan of oatmeal, sawdust, or grain. First, however, expose the food for a few days in the pan without setting the trap, until a rodent readily takes it.

6. *Dispose of dead rodents properly.* In getting rid of dead rats and mice, it is important to avoid any contact with the parasites they commonly carry, such as fleas, lice, and mites. These are the real carriers of the diseases we can get from rodents. Use a long-handled shovel, long-handled tongs, or two sticks, and wear long rubber gloves. If you don't have any of these devices available, slip your hand into a plastic bag that is long enough to reach your elbow and big enough to hold the dead animal, pick up the rodent with the same hand, invert the bag over your catch, and tie the end securely. If you have caught a sizable number of animals, you

can burn them (if that is legal in your community) or bury them deep enough so that other animals can't dig them up. Should you catch only one or two, you can discard them with the garbage.

If a rodent dies within a wall void, you will have to deal with the odor sooner or later. You might consider taking a long vacation, but there are less expensive ways to solve this problem. Fans, including the vent fan in a room air conditioner, will help to drive out the stench. You can also use a spray or mist, or a cotton wick dipped in a bowl of one of the following (you may need to check with a pharmacist or supplier of chemicals for some of these):

- Bactine
- Isobornyl acetate (a pine-scented liquid that is used as an ingredient in perfumes, cleaning products, and other substances)
- Neutroleum alpha
- Quarternary ammonium compounds
- Styamine 1622

Live Rodent Panic!

It is possible that a rodent will be caught but not killed by a trap you set. If this happens to you:

1. *Allow yourself one gasp.*

2. *Get out the dustpan and brush, and sweep both the trap and the victim into an empty paper bag. Fold the top of the bag down as far as it will go.*

3. *Submerge the bag in a bucket of water and hold it down with a brick, a flowerpot, or a good-sized rock.*

4. *As an alternative to 3, above, set the paper bag in a larger, clean plastic bag and put it in the freezer. After two or three hours, move it to a refuse can.*

If you find a mouse in a cereal box, you can treat it in the same way, putting the whole box into the plastic bag, and either drowning it or freezing it. A sharp blow to the base of the skull provides quick euthanasia. Either of these methods is more humane than putting a live animal into a trash barrel to be crushed in the garbage truck's compactor or buried alive in a sanitary landfill. However, if you feel too squeamish to use either of these methods, you might recruit a kindly, stouthearted friend or neighbor for the task.

Essential oil of pine, peppermint, wintergreen, or anise; formalin (a formaldehyde-based solution); and activated charcoal are also effective odor masks.

If you can locate the carcass inside the wall—do flies gather there? does your dog keep sniffing there?—drill a small hole through the wall a few inches above the floor and as close to the odor source as possible. Using a narrow tube, pour in one of the masking agents listed above. This is usually the fastest way to clear out a stench.

Finally, following are some suggestions and observations on the use of snap traps from the U.S. Public Health Source, which has had tremendous experience in rat control:

• A dirty trap may be more effective than a clean one. A trap that has already caught a rat will be even more attractive to its fellows.

• More rats are trapped on the first night than any other night. After three or four nights, the catch will drop to zero, making it imperative that you lay enough traps initially.

• If rats are getting in through a hole in the floor, set several traps in a semicircle two or three feet from the opening.

• A light layer of flour (or talc) spread along rodent runways will often show you the best places to lay traps.

• For roof rats, which are much harder to trap than their Norwegian cousins, try holding traps in place with bungee cords.

• Every few days, move the traps about two feet to a new location.

Glue Boards

Glue boards—traps smeared with a sticky substance—will catch mice. However, unless two of its feet are caught, a rat can pull itself free. Furthermore, if you are squeamish about disposing of a live victim, its distress cries will disturb you until the animal dies. If you decide to use glue boards, set them inside bait stations, where they won't get dusty or stepped on. For hanging near ceilings or on rafters, bend them into a cylinder, tape them closed and then tape them to pipes near ceilings or narrow ledges. These traps are being used successfully today in food facilities, where they catch thousands of mice in a matter of days. However, they are considered less humane than snap traps.

Rodenticides

Various rodenticides, some more toxic than others, are sold to control rats and mice in homes. *All rodenticides are dangerous enough to cause death. Despite advertising claims, no known rodenticide is completely safe. Poison is poison.* Predators of rodents, such as cats, owls, and hawks—our allies in this endless war—are easily killed by these toxins. In addition, the widespread use of warfarin, a substance that interferes with blood clotting, has led to extensive resistance to this chemical. In some parts of the United States and Canada, scientists have observed 75-percent resistance in house mice populations. In other words, like so many other chemical controls, warfarin is losing its effectiveness.

Rampage and other commercial products that are based on Vitamin D_3 cause calcification of rodents' breathing systems, so that they die two to four days after a lethal dose. With any poison, however, pests may die within the walls of your home, leaving you with the unsavory job of dealing with dead-rodent odor (see page 91).

If someone offers you a yellow paste that resembles peanut butter and guarantees that it will kill rats and mice, *refuse it*. This is yellow phosphorus, a poison so powerful that it can kill a small child who walks over it barefoot. And there is *no* antidote for this pesticide. It may be called J-O Paste, Paste Electrica, Rough-on-Rats, or Stearns Electric Paste. Although it is illegal for use as a pesticide in the United States, yellow phosphorus still comes into this country by way of Latin America.

Why Not Just Call in the Cats?

Wouldn't it be simpler just to get a cat and turn it loose? Not really. Professional pest control operators think that a cat is worthless against an infestation of rats. Says one New York City rat-control officer, "You put a cat in a cellar with a heavy rat infestation and you've got yourself one very dead cat, believe me."

Some dogs—silky terriers or toy fox terriers, for instance—may be good rat-catchers, but, generally, a cat or dog will get the worst of it in any dust-up with a ferocious buck rat. In packs, rats can gang up and best even a wild boar or a tiger.

Studies have shown that in residential areas, cats kill only about 20 percent of the number of rats that must die every year to maintain a stable rodent population. On farms, cats kill enough rats to prevent an upsurge in rodent population. But in neither type of location do cats exterminate the rats.

The fact is that cats and dogs may actually draw rats to an area. A rat can live very comfortably under a doghouse, eating and drinking the pet's food and water while the rightful resident is asleep. In general, cats and dogs just keep the rodents out of sight, where the pets cannot reach them. However, pets may be useful in keeping an area free of rodents *after* the vermin have been killed by other means.

And don't count on owls or hawks to do the job of rodent eradication, either. Along with foxes, snakes, humans, cats, dogs, and parasites, such rodent predators cut down the pest population only temporarily. The weight of scientific evidence indicates that the number of predators is actually regulated by the number of their prey, and not the other way around.

Once you have cut off the rodents' food supplies, destroyed their harborages, and killed those that had set up housekeeping with you, you can concentrate on barricading your home against them, using the methods outlined under Shut Them Out, earlier in this chapter, to prevent their ever becoming a problem again.

Future Controls

Environmental researchers are working along several tracks to sharpen our weapons against domestic rodents. One route being studied is controlling rodent fertility. Another is lacing baited poisons with pheromones (sex lures) to make them more attractive to these wary animals. A third is biosonics—that is, mimickings or recordings of rodent distress sounds, played back with good acoustics and at low intensity to lure the animals into traps. The practical use of any of these approaches is most likely years down the road, however.

Until something is proven to be better, we will have to continue to depend on sanitation, rodent-proofed buildings, and well-laid traps to keep rodents—adaptable, wily, and prolific as they are—from overrunning our homes and farms.

6. The Fearsome Fly

Once upon a time, a tailor sewing in the window of his shop was bothered by a swarm of flies. Whacking about with a piece of cloth, he killed seven of the buzzers at one blow. So proud was he of this feat that he stitched his belt with the words "seven at one blow," and left his shop forever to seek his fortune. Parlaying his skill as a fly-slayer into a reputation for great bravery, he won the hand of a princess along with her father's kingdom.

The plucky tailor well deserved his good fortune, for he had destroyed seven of humankind's most dangerous enemies.

TYPES OF FLIES

There are many different species and types of flies. They may seem similar at first, but if you look closely, you can see that there are differences.

Common Houseflies

Common houseflies are probably the best known fly species. They need no particular introduction.

Little (or Lesser) Houseflies

This species physically resembles the common housefly, although it is somewhat smaller. It is perhaps easiest to distinguish by the fact that it tends to flit about more than the common housefly does. Little houseflies are not the young of common houseflies, but members of another species.

Little houseflies tend to appear earlier in the spring than common houseflies, and to become inactive or even disappear in hot weather. They are attracted by animal droppings and other decaying organic matter. They also like honeydew, the syrupy substance secreted by aphids and other garden insects. The syrup eventually becomes moldy and turns a sooty black.

Black Garbage Flies

As the name suggests, these are black in color. They are a bit smaller than common houseflies.

Blowflies

In general, blowflies are slightly larger than common houseflies. They tend to be shiny and metallic-colored. Bluebottle flies and greenbottle flies are types of blowflies.

Cluster Flies

Flies of this species are gray (they may be varying shades of gray). They are slightly larger than common houseflies.

Although not disease carriers, cluster flies can certainly destroy a householder's peace of mind. Widespread throughout Canada and the United States (except along the Mexican border), these insects can appear suddenly on warm winter days. They breed in moist, earthworm-rich soil, and are dormant in winter until a sudden warm spell awakens them. Then they invade homes in huge numbers through small cracks. Once inside, they gather in closets, in clothing, on furniture, on dishes, and around lighted lamps. Swatting them leaves ugly stains on furniture and walls, and the dead flies can draw other pests that feed on the remains.

Drain Flies

If a small, brownish-gray hairy monster emerges from your bathroom drain, don't panic. You're not in a horror film gone Lilliputian. This is a drain, or moth, fly. They breed by the millions on sludge in municipal sewage treatment facilities, and some stray far from their origins. These pests are attracted by gelatinous accumulations in sewers, septic tanks, tree holes, and rain barrels. They also like damp, dirty garbage pails and moist birds' nests.

Fruit (Vinegar) Flies

These tiny flies breed in rotting or overripe fruit, uncooked food, and excrement. Their ability to reproduce quickly makes them an important tool in biological and genetic research. It also makes them potentially serious pests.

Hover Flies

Many of these flies mimic bees or wasps in their coloring and hovering behavior.

Midges (Gnats)

As more people move to previously undeveloped lands of the desert, mountains, and seashore, midges (also called gnats)—which in more developed areas usually annoy us only when we're outdoors—can be a serious problem. Some of these pests can cause serious allergic reactions, even the swelling of an entire arm or leg, as well as fever. If they swarm against an automobile radiator, the engine can overheat. Massed on a car windshield, they can blind the driver and cause an accident. They are such a nuisance in Florida and the Caribbean that they have caused drops in tourism. They are sometimes a problem on golf courses and in housing developments with manmade lakes, where warm, organic-rich stagnant water provides ideal breeding sites.

Stable Flies

These creatures resemble common houseflies. However, if you look at one closely, you will notice that its head tapers to a more slender point.

False Stable Flies

The false stable fly also generally resembles the common housefly, although it tends to be slightly larger and wider. It is gray in color.

Tachinid Flies

These creatures generally resemble heavily bristled houseflies. Unlike houseflies, however, they come in many different colors.

For illustrations of some common types of flies, see Figure 6.1.

WHAT FLIES EAT

Flies can feed on bits of food (whether fresh or rotting); human, animal, or plant waste; and even sweat. Though some species may favor one type of food over another if given a choice, flies are highly adaptable and can survive on incredibly varied diets. All they need is organic matter of some kind—which covers such a range of things it would be impossible to name them all here.

THE HAZARDS OF FLIES

Along with their cousins the mosquitoes, flies have caused more human and animal illness and death than any other animal on earth. Although the evidence in some cases is circumstantial, the common housefly, one of several kinds of filth flies (so called because they pick up and transmit germs from sewage, garbage, excrement, decomposing bodies, and other such sources), is the probable carrier of

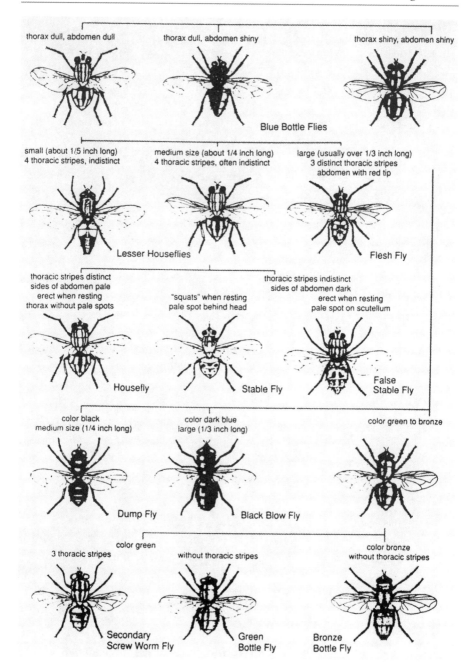

Figure 6.1 A Pictorial Key to Common Domestic Flies

Source: Files of Public Health Importance and Their Control (Atlanta: Centers for Disease Control, Homestudy Course 3013-6, Manual 5, 1982).

more than sixty-five different human and animal diseases. Houseflies have been linked to outbreaks of cholera, anthrax, dysentery, tuberculosis, typhoid fever, plague, yaws (a chronic bacterial disease most common in tropical areas), leprosy, tapeworm, gonorrhea, and polio. According to the U.S. Centers for Disease Control and Prevention, the common housefly is a greater danger to humans than any other species because of its close association with us, its filthy habits, and its ability to transmit germs.

Not all types of flies are dangerous, however. Some are even considered helpful. Among these are hover, or syrphid, flies, and tachinid flies. (More information on these species may be found in Chapter 12.)

However, blowflies, stable flies, false stable flies, houseflies, and little houseflies should be kept as far from our homes as possible. Blowflies, either bluebottle or greenbottle, that buzz around a lamp at night or settle on a kitchen counter can fly up to twenty-eight miles from their breeding grounds in garbage dumps or slaughterhouses to contaminate our food. They often lay their eggs on fresh meat minutes after it is exposed to the air. If swallowed, the eggs and/or larvae can make us very sick.

Midges and fruit flies—both true flies—are also a danger. So tiny that they can pass through ordinary window screening, some of these minicreatures can give painful, long-lasting bites. Some are bloodsuckers. Entomologists have traced outbreaks of conjunctivitis (pinkeye) and trachoma (a chronic form of conjunctivitis) to so-called eye gnats, which rupture the eye's protective membrane. If left untreated, trachoma leads to blindness.

HOW FLIES MAKE THEMSELVES AT HOME

Many species of flies feed and lay their eggs in cesspools, stagnant water, dung heaps, garbage cans, and Dumpsters, as well as any place else there is rotting animal or vegetable matter. Bacteria thus enter their bodies and are also trapped on their leg hairs. In her one month of life, a female housefly lays from 350 to 900 eggs. Her offspring, or maggots, feed on the filth from four days to two weeks. After pupating (resting in an outer case) for less than a week, a fully developed adult fly emerges, and the cycle continues.

When they get into our homes, sample our food, walk on our dishes, or sip our sweat, houseflies contaminate us with their eggs, larvae, bacteria, and excreta. Outbreaks of diarrhea have been found to correlate with the warm-weather surge in the fly population. So don't shrug off those black buzzers hovering in your entranceway or around your patio.

Although chemists have developed a whole battery of powerful sprays to control pest flies, in a few short generations (and there may be ten generations a summer) flies often become immune to the chemicals. A housefly that is heavily dusted with DDT powder can go on calmly feeding on some rotting food without any bother.

Not long ago, health authorities felt that the fly problem in the United States was not serious, but now they are changing their minds. For the last century, our population has been making deeper and deeper inroads into what formerly was the countryside. At the same time, as our numbers grow, we are consuming vastly more meat and poultry than ever before, and these have to be produced somewhere. Instead of being raised (as they once were) on widely scattered farms— where the insects drawn to their droppings were soon eaten by predators—cattle, hogs, and chickens are now gathered in mass feeding facilities, where important insect-eaters have been eliminated. Unless they are meticulously managed, these "meat factories" quickly become "pest factories."

Your new suburban home may be only a short distance from a cattle feed lot or poultry ranch. In addition, the tender new lawn you have been conscientiously fertilizing will add to your fly problem. This urban/rural interface is happening with increasing frequency. You may have to cope with flies for several years, until neighborhood lawns are well established and additional development pushes the livestock farther away.

Homeowners who have only recently moved to a new development from a city or older suburb may find the flies more bothersome than do people who have lived in the area for a while. Where countryside and farm are giving way to suburbs, there will be more flies, rodents, snakes, and wasps than you are used to— they come with the territory. Until public sanitation is well established, you will need to have more tolerance for small wildlife, and to monitor the sanitation rigorously around your own home.

SHUT THEM OUT

Fly-proofing a house is fairly easy, and most of the controls are mechanical.

Screens and Fans

Tightly fitting window screens (16-mesh is the standard) and outward-opening, self-closing screen doors are as foolproof a fly barricade as anything yet invented. Whether metal or plastic, the screening must be in good repair and firmly secured to its frame.

If your door to the outside is used a great deal, install a ceiling fan above it that moves at least 1,500 cubic feet of air downward every minute. This is a ploy you will find in many butcher shops and restaurants, where doors are open as much as they are closed. If you have a covered patio and enjoy outdoor meals, set an overhead fan there, too.

Repellents

N,N-diethyl-meta-toluamide (DEET), commonly used to repel mosquitoes and ticks, will deter flies as well. (For a detailed discussion of DEET, see page 125 in Chapter 7.)

Are you interested in experimenting with herbal controls? Try growing tansy near your kitchen door or wherever flies tend to cluster. Though this is not a substitute for meticulous sanitation, rural people once found the plant to be a good fly repellent. Other effective natural fly repellents are oil of cloves and pine oil.

Incidentally, if a friend suggests that you plant a castor bean tree near your kitchen to repel flies, ignore this advice. The castor bean is the source of ricin, a poison so deadly it has been used in political assassination; less than 1 milligram can kill a person. In January 2003, London police reported finding traces of it in the apartment of suspected terrorists. Every year, several children are poisoned by eating castor bean seeds.

Other Controls

There are a number of pest fly species that pose health dangers and have habits different from those of common houseflies. If you have a problem with any of the following, you may need to adopt other strategies to control them.

Fruit (Vinegar) Flies

If you make your own wine or beer, you have to be meticulous about keeping the containers, lids, and utensils clean inside and out. Also eliminate any special attractors you may have out in the open, including:

- Containers of cider, vinegar, beer, or wine whose contents have spilled.
- Ripe fruits and vegetables.
- Juices, ketchup, and sour milk.
- Souring scrub water and mops.
- Dirty garbage cans.

In addition, take measures to eliminate breeding and feeding sites and deny fruit flies entry to your home, including the following:

- Consider changing to finer-meshed door and window screens.
- Clean up any fruit peelings, spilled juice, or milk under your refrigerator or stove.

- Store ketchup in the refrigerator.

- Thoroughly rinse all dairy cartons, juice cans, and ketchup bottles before discarding or recycling them.

- After washing a floor, wipe it dry with a towel or rag. Don't let water stand in any crevices.

- After cleaning, rinse your mop thoroughly and place it in direct sunlight for a few hours. If the weather is rainy or cold, keep it in an indoor area where you dry clothes.

Drain (Moth) Flies

To discourage drain moths, take the following measures:

- Close up any nearby tree holes.

- Discourage nest-building birds.

- Keep rain barrels covered in dry weather.

- Drain all standing water.

- Handle garbage hygienically. (For ways to do this, see information on outdoor sanitation on page 106.)

- Occasionally, scour the insides of drains with a swab of paper wrapped in waxed paper.

- About once a month, pour a quart of boiling water down each drain (*not* the garbage disposal) to prevent a buildup of goo. This keeps the drains clear and adds no harmful substances to your plumbing.

Little Houseflies

The best approach to controlling these insects is to eliminate breeding and feeding places:

- Monitor food waste carefully.

- Eliminate honeydew-producing garden pests. These include aphids, mealybugs, scale insects, and whitefly. (For more information on garden pests, see Chapter 12.)

Midges

If you must go outdoors when these flies are swarming, use a repellent such as DEET. (See page 125 for a detailed discussion of this repellent.)

If you are planning a vacation in the mountains or desert or at the seashore, ask health authorities in the region you will be visiting when gnats and midges are usually active. Then plan accordingly.

STARVE THEM OUT

The secret to fly control can be summed up in one word—sanitation. Your key word here is SOS—"Seek Out the Source." Every adult fly near your home represents a breakdown of sanitation procedures. To a great extent, we make our own fly problems with our careless handling of food waste and other decaying organic matter. Some years ago, two scientists undertook to count the number of flies produced by storing refuse in defective garbage cans. These unsung heroes found out that, in summer, the average barrel bred over 1,000 flies a week. One barrel (which must have belonged to a *first-class* slob) produced nearly 20,000 in one week.

If you have a young adult child who is beginning to keep house in a college apartment, you may get a phone call like one I received a while ago:

"Mom," said this future scientist, "there are little white worms in my garbage. What should I do about them?"

"They're maggots of flies. Better sharpen your housekeeping," I advised.

Now's the time for you to begin a cram course, Kitchen Sanitation 101, for this novice.

First, how you handle food waste is critical. Like all insects, flies have a keen sense of smell. Though drawn by almost any food odor, they don't go just for the pleasant ones; a stench that turns your stomach sets this pest's mouth watering. One of your major goals in handling food waste must be to control the odors of decay. Making a habit of the simple steps in the following sections will do just that.

Indoors

❑ Rinse clean all cans, jars, dairy cartons, and meat, fish, or poultry wrappings, and drain them well before throwing them out.

❑ Keep a plastic bag in your kitchen garbage pail for food waste. Drain the scraps well, and twist the bag closed each time you put refuse into it. At the end of the day, close the bag with a twist-tie or rubber band and take it outside—to the incinerator if you live in an apartment; to a large, strong, closely covered garbage can if you live in a single-family house.

❑ Handle disposable diapers hygienically. Unless you dispose of them carefully, those cushiony little pants can turn a family's garbage cans into disease-breeding toilets.

Community Sanitary Standards Recommended by the Centers for Disease Control

Adequate community sanitation and refuse control are essential to any program aimed at eliminating flies. What does adequate mean here? How does your community measure up? What happens to refuse after it leaves your control? How can you know what good community sanitation is?

The following standards for safe refuse handling have been established by the U.S. Centers for Disease Control and Prevention. These recommendations should be followed closely by every community official concerned with citizen health.

PRIVIES AND SEPTIC TANKS

There are still places in the United States without indoor flush toilets. Many are in substandard rural and semirural homes, or are public facilities in poorly developed small communities or campgrounds. Properly installed and maintained septic tanks can be a preliminary measure in places far from a modern sewage system, but they are acceptable only as a temporary measure. As soon as it can be arranged, the household or public recreation area using a cesspool should be hooked into an up-to-date municipal sewerage network.

For homes or recreation facilities in remote areas, composting toilets may be the best long-term solution. There are several types of these currently on the market. They are designed both to be odor free and to produce usable fertilizer. Check with your local health authorities to see which type of installation is acceptable in your area.

INDUSTRIAL WASTE

City and county health and sanitation officials should continuously monitor any feedlots, slaughterhouses, packing plants, canneries, and feed mills within their jurisdiction. Waste recycling facilities and landfills should also be inspected regularly. Such facilities can be prime sources of disease-bearing pests.

REFUSE COLLECTION

"The collection system," say the Centers for Disease Control and Prevention, "must be designed for the improvement of sanitation and not for the convenience of the collection agency." Refuse should be collected from residences twice a week in summer, and daily from restaurants, hotels, and food markets. Collectors should be careful not to spill garbage or damage the containers. Their trucks should be closed, of the packer type, and kept clean.

FINAL DISPOSAL

Some towns and cities in the United States still drop their garbage on vacant acreage at what the officials think is a safe distance from the community. If you think you are safe because your town burns its refuse at a dump "way out in the country," you're wrong. Houseflies have been found 100 miles out at sea, blown there by the wind.

A burning dump is a menace to a community, attracting filth insects and verminous rodents. In addition, exploding glass and aerosol containers can injure anyone in the vicinity. Proper disposal of a city's garbage can be done in any of three ways: individual garbage disposals for each household, incinerators, and sanitary landfills.

Garbage Disposals

When you use a garbage disposal, your pulverized food waste becomes part of the city's sewage, and is treated at its reclamation plant. To prevent the occasional stuffed sink (and why does this always happen when you're having a dinner party?), make sure all the garbage is completely flushed away by running a strong stream of water for a full minute after you shut off the appliance.

Incinerators

Whether for a whole city or a single apartment building, incinerators should operate at temperatures from 1,400°F to 2,000°F. A defective installation will merely char the garbage, barbecuing it nicely for any resident rats and cockroaches. Several European cities get double value from their municipal incinerators by using them to produce electricity. We Americans could benefit from their example.

Sanitary Landfills

In a sanitary landfill, compacted garbage is dumped on swampy or otherwise useless land some distance from the city. There it is covered *every day* with at least six inches of earth. When the parcel is completely filled, it is buried under a minimum of two feet of soil. Some cities find it sufficient to shred, mill, or pulverize the garbage to destroy maggots before dumping it, and use no soil cover. As a side benefit, a sanitary landfill reclaims otherwise unusable land and makes it habitable.

If public health officials, elected or appointed, are permitting public or private fly-breeders to menace your community's health, become a fly yourself—a gadfly—and nag them until the hazards are eliminated.

Outdoors

❏ Keep outdoor garbage cans and bins in good repair. Metal barrels should have recessed bottoms to prevent rusting.

❏ Make sure garbage containers have sound, tight lids. It does no good to set garbage outside if the bin's lid is missing or even just cracked. A fly can squeeze through a slit to get at a moist, decaying meal. If you cannot replace a cracked lid with a snug one (some dealers sell lids alone), get a new container. Tight covers on sturdy barrels can reduce the local fly population by 90 percent if the garbage placed in them is well wrapped.

❏ Make sure to store outdoor garbage containers properly. They should be stored on a sanitary rack that is at least eighteen inches off the ground to help prevent bacteria-favoring moisture from collecting underneath. (See Figures 5.10, 5.11, and 5.12 on page 86.)

❏ Never use trash bags alone as garbage receptacles. Cats, dogs, rats, racoons, and—in the Far West, coyotes—can quickly tear them open.

❏ After an outdoor barbecue, be sure to include the grill in your cleanup so that bits of meat or fish don't give flies their own cookout.

❏ Be sure to pick up dropped fruit from trees.

❏ To prevent lawn-bred flies, spread manure in a thin layer so that any eggs and maggots laid in it will die of heat, cold, and/or dryness. Gather lawn clippings in a plastic bag, and put the bag out for the collectors.Anyone who has ever fertilized a lawn or been in the neighborhood when manure is being spread knows how flies swarm to it, but even without fertilizer, a lawn can be a fly source. Decomposing trimmings may also add to your fly problem. Not long ago, an entomologist raised twenty-five filth flies from grass clippings caked on the blade guard of a power mower.

❏ Dispose of pet droppings promptly and properly. Having a fine dog can add much to your life, but unless you clean up after it, the local pest population will soon outrun your ability to curb it. The same is true of that rare being, a sloppy cat. And you will not only be coping with more flies, but with a lively bunch of cockroaches, too. There are sanitary ways to deal with pet droppings:

   • Bury the droppings. Run a sprinkler for several hours in an unpaved part of your yard, and then dig a hole two or three feet deep. Drop in each day's cleanup, and cover it with a light layer of the excavated dirt. Be sure, however, that your dog is free of worms, as the eggs can live a long time in the ground and be picked up by anyone walking there barefoot.

- Install a dog latrine. Set a dog latrine (commercially available at better-stocked pet stores and home centers) in the ground. As long as it is properly maintained, this should work if your dog is not too large.

- Dispose of droppings in a plastic bag, and then set the bag in a tightly sealed garbage can.

- Flush the droppings down the toilet.

- If you cannot dispose of droppings right away, sprinkle them with a thin layer of sand or sawdust to minimize stench and moisture.

- Confine your pet. Thoughtful people don't let their animals set up fly-breeders on neighbors' lawns or public property. In New York City, this became such a serious problem at one point that sanitation police began ticketing people whose animals fouled the city's streets. Other cities have followed New York's example. In Pasadena, dog owners routinely carry a small plastic bag and a scoop when walking their pets, and some parks in the San Francisco Bay area thoughtfully provide bags for pet owners who don't have their own on hand.

❏ If you give your dog a bone to gnaw outdoors, wrap it up in plastic after your pet has finished with it and throw it out. Dog bones also can draw flies. (Bones left outdoors can attract rats as well.)

❏ If you yourself are always meticulous, but there are still too many flies in your area, alert your state's environmental protection agency. Your community may still be sending organic waste to an open dump. With technical and financial help, the appropriate agency can guide your local authorities in setting up a sanitary land-fill. You yourself can check the neighborhood for rotting organic matter or stagnant water nearby. If the problem is severe, call your property owners' association or health department. They can contact careless individuals or alert the whole neigh-borhood to the potential for disease. Officials can also have standing water on pub-lic property drained or order empty lots being used as dumps or pet latrines to be cleaned up and fenced off.

❏ If you are plagued with cluster flies, keep your grass mowed to a height of no more than two or three inches to let sunlight in and dry out the soil. You may be sorry to do this if you are a devoted gardener who cherishes every earthworm, but letting the soil dry out kills the earthworms and denies the cluster fly young the food they need. Also make sure that all small cracks on your home's exterior are filled and that weatherstripping around doors and windows is snug.

WIPE THEM OUT

There are a number of strategies you can use to wipe out flies, of whatever type.

Traps, Swatters, and Lights

As long as people generate waste, flies will be with us. Your goal is to keep them to a minimum.

An outdoor fly trap can do a great job of controlling flies. Commercially available traps can cost as much as thirty-five or forty dollars each. How well a trap works for you depends on where you set it and how often you replenish the bait. (Incidentally, be wary of the bait recommended for a commercial trap. If raw liver is suggested, ignore it. When I tried this, after a week the nauseating stench attracted a coyote, which killed our beloved cat.)

All fly traps have to be emptied from time to time, an unpleasant job at best. Traps of various designs come and go on the market, so you may be better off making them yourself. By using the following simple measures, you can greatly cut down the number of flies in and around your home.

❏ Set out an inexpensive trap made of a baited jar covered by a lid-like strainer. To avoid the risk of shattering, use a plastic jar.

❏ A suggestion from *Mother Earth News:* Shape a piece of paper into a cone, and insert it into the neck of a baited jar.

❏ You can still buy flypaper, but some public health authorities consider it ineffective. Do not hang pesticide-impregnated strips indoors or near a spot where people gather. These strips constantly release toxic vapors into the air. Many of them contain chlorpyrifos, an organophosphate insecticide that is a known carcinogen. The U.S. Environmental Protection Agency (EPA) has now banned it, but stocks may still be on store shelves..

❏ For the occasional invader, an expertly wielded fly swatter is as potent as any aerosol spray, and a lot safer. A quick sudsing in hot water after the kill, followed by sun- or air-drying, takes care of any contamination on the swatter. If you can get the hang of using a rolled-up newspaper or magazine for a swatter (I confess I never have), you've got the ideal weapon—effective, sanitary, and disposable.

❏ To get rid of flies in the daytime, darken the room they are in and open the door to the light—the brighter, the better. Like many winged insects, flies are drawn to light. If you open the screen door, they will soon fly out. If the screen door is closed, the flies are sitting ducks. If you aren't too good with a swatter, spray the insects with rubbing alcohol for a quick kill and few escapes. (Be careful with this method, though, because alcohol can mar paint if it drips.)

❏ If flies get in at night, try this: Before you go to bed, close all of your draperies and window shades, leaving a narrow slit at one of them. In the morning, the flies, torpid in the cool air and dozing on the bare, bright strip of glass, will be easy prey.

❏ If despite your best preventive measures, cluster flies invade your home, you can install light traps for flying insects. To be effective these should be installed between eighteen inches and five feet above the floor. Those hung any higher will not catch flies. Try to avoid placing them near windows, because they can attract other insects; they will be most effective in a darkened room.

Chemical Fly Control

If you have an infestation of drain flies, *do not* pour insecticide down the drain. This is illegal in many communities, and is also hazardous to everyone in the vicinity. Instead, pour boiling water down the drain periodically.

If you have a serious problem with midges, it is best for health officials to apply suitable insecticides.

Nature's Fly Controllers

If you have ever had occasion to use a country privy, were you surprised that, despite the odor, there were few, if any, flies present? Spider webs veiling the window and wasps' nests in high corners were signs that the local sanitation crew was on the job.

Spiders

Flies soon become immune to the chemist's most powerful toxins, but in millions of years on earth, no fly has ever developed immunity to any spider. You may be less fond of spiders than you are of flies, but most of the small spinners are harmless. (Ways to control unwanted spiders will be covered in Chapter 13.) In general, spiders—even the poisonous ones—rarely bite people.

Although spiders prey on various insect pests, their favorite dish seems to be flies. So when a harmless spinner sets up housekeeping in an out-of-the-way corner of your property, try to resist the urge to sweep away the web and step on its builder. Chances are that it is posing no danger to you. Your tolerance will be well repaid, for each of these skillful weavers can destroy as many as 2,000 pests in its short life.

If your washing machine is positioned under a window in an otherwise dark garage or basement, you may regularly find spider webs at the window. Let them stay. Drawn by the light and moisture of the laundering, flies buzz against the window, where a watchful spider soon adds them to its larder. When the web becomes too large and gets in the way, brush half of it away. In a matter of weeks, the spider (or its children) will rebuild it and continue its patrol.

To be certain that the spider working your territory is not a potential danger to you, put on a pair of heavy work gloves, scoop the animal into a small box or jar, take it to a local arboretum, natural history museum, or community college's biology department for identification.

Wasps

A more temperamental but possibly even more effective ally in controlling flies is the wasp. Of course, you don't want to get too chummy with wasps or let them indoors. However, wasps are voracious pest predators that feed their young on many different kinds of insects. A century ago, iron-nerved Americans actually hung wasps' nests near their stable doors to reduce the number of flies. A colony of yellow jackets or paper wasps nesting in a little-used corner of your yard will provide more thorough and longer-lasting pest control than any chemical. (For ways to control wasps themselves, see Chapter 13.)

Future Controls

Despite the fact that there are many species of pest flies, the one of most concern remains the common housefly. Scientists continue to seek ways to keep their numbers within safer limits. Absolute safety would be no flies at all, but this is impossible. Since chemicals are clearly not the answer to fly eradication, researchers are looking into biological controls.

Nonpest fly larvae that compete with housefly young for dung, and larval parasites that destroy hatchlings, are under study. Both seem to hold promise. There are also several kinds of parasitic beetles that feed on fly larvae. Harmless to people and other animals, they feed voraciously on fly maggots. They are extremely effective in chicken houses, cattle pens, and places where manure accumulates.

The release of sterile male flies has had spectacular success east of the Mississippi River against the once-devastating screwworm fly. This insect's maggot can infest 20 percent of a cattle herd, kill 20 percent of those infested, and be ingested by humans who eat meat that comes from an affected animal. Infestations in western states will end completely when scientists can figure out how to prevent reintroduction of the screwworm fly from Mexico.

Researchers are also looking into types of bacteria and fungi that feed on maggots, preventing them from developing into pestiferous adults.

Progress in biological control will take time. Until we learn how to disrupt the housefly's breeding cycle with reliable techniques and how to apply those methods throughout a state or even the nation, rigorous monitoring of animal, human, and food waste in our homes and communities remains our best means of control.

PART THREE

Pests of
the Body

7. The Mosquito— A Deadly Nuisance

A mosquito that had been frozen in amber for 30 million years started renowned writer Michael Crichton on a fiction-writer's favorite game, "What if . . . ?" What if that tiny fly had bitten a dinosaur before it became trapped and the insect still had some of the huge lizard's DNA in its body? Could a geneticist replicate the dinosaur and bring it to life? In pursuing his thought, Crichton, along with motion picture producer Steven Spielberg, gave us the blockbuster *Jurassic Park.* But there is nothing entertaining about the other "gifts" the mosquito has brought. Over the centuries, mosquitoes have caused humanity more grief than any other insect (with the possible exception of houseflies), even more than plague-bearing rat fleas that killed three-fourths of the people in Europe and Asia 600 years ago. Actually a specialized type of fly, this tiny animal is a worldwide problem. Mosquitoes may inflict more misery on humans than any other type of insect.

Individual mosquitoes are almost weightless—10,000 of them together can tip the scales at less than an ounce. They are leggy creatures with long, narrow wings, and are rather dainty if you can take the time before slapping to look at one carefully. Some are even beautiful, with splashes of silver gleaming on dark bodies or black and white legs.

Despite their delicate appearance, mosquitoes are without doubt one of the most dangerous animals on earth, carrying deadly diseases to millions of human beings every year.

As already mentioned, mosquitoes are actually flies; *mosquito* is the Spanish word for "little fly." Depending on which expert you consult, there are 3,000 to 4,000 different species of these insects, with 400 to 500 of them capable of transmitting diseases. Both sexes feed on flower nectars; however, the males never bite humans or other animals. The females, however, must have a meal of blood before they can lay fertile eggs.

Some species feed during the day, some at dusk, others at night. In warm climates, they are abundant all year. In cooler regions, they can appear as early as

February, when melting snows form the quiet pools that the females seek to shelter their eggs.

Mosquitoes have four life stages: egg, larva, pupa, and adult. The eggs are laid on or near water, or in places where water will eventually cover them. The larvae and pupae live underwater but have breathing tubes to fresh air. The adults live on dry land. (See Figure 7.1).

All mosquitoes require still water on which to drop their eggs. Depending on the species, the site may be:

• A place with permanently stagnant water, such as a lake, pond, or salt marsh.

• A place with temporarily still water, such as a ditch, clogged stream, irrigation canal, or container near a home that can collect water.

• Flood water. This can be the result of a natural cataclysm such as a hurricane or simply more rainfall than natural drainage can handle. Following Hurricane Mitch, which hit Honduras in 1998, that country suffered 30,000 cases of malaria. In Yemen, in 1999, after a season of unusually heavy rainfall, 2,000 people died of the disease.

All three kinds of the water sites mentioned above abound in and near cities and in the countryside. If a mosquito lays eggs in a small puddle that dries up before the eggs hatch, the larvae will emerge almost as soon as water accumulates there again—even if (for some rare varieties) it is *five or six years later.*

The wiggling larval stage may last a week; then the animals become relatively inactive pupae. Three or four days later, the adults appear, ready to mate and, if female, hungry for a blood meal. Though some species can fly many miles, some of the most common species, like the house mosquito, will travel only to the nearest blood feeding.

THE HAZARDS OF MOSQUITOES

Despite their delicate appearance, mosquitoes are without doubt one of the most dangerous animals on Earth, carrying deadly diseases to millions of human beings every year.

Historians say that mosquitoes were partly responsible for the fall of Rome. They bred in swamps near the city and infected many of the city's inhabitants with malaria. In the late nineteenth century, these dainty insects brought the construction of the Panama Canal to a complete halt as the French builders, one after the other, succumbed to yellow fever and malaria, both of which mosquito-borne diseases raged on the isthmus of Panama. A century after that, a moon shot was imperiled when Louisiana rocket builders threatened to quit because day-biting mosquitoes were attacking their children. And magnificent Yosemite National

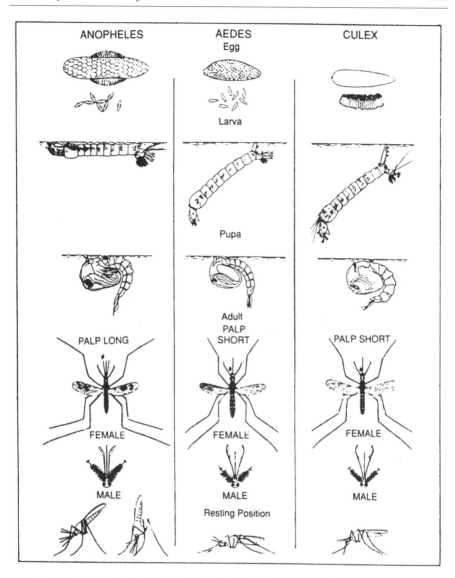

Figure 7.1 Life Cycles of Three Species of Disease-Bearing Mosquitoes
Source: Insect and Rodent Control: Repairs and Utilities (Washington,DC: U.S. War Department, Technical Manual TM5 632, October 1945).

Park was uninhabitable until health authorities cleared out the area's dense clouds of mosquitoes. Even today, park rangers tell visitors, "Don't feed the deer. Come in the spring and feed the mosquitoes." In Alaska boosters say, "Come to Alaska . . . a billion mosquitoes can't be wrong." Minnesotans like to refer to the mosquito as the state bird.

In all fifty U.S. states, most public efforts against pests—80 to 90 percent—are aimed at suppressing mosquitoes, which are carriers of malaria, yellow fever, dengue fever, and many forms of encephalitis (inflammation of the brain). Even our dogs don't escape, for mosquitoes can give them heartworm.

The amount of money spent on suppressing mosquitoes is astronomical. The cost of materials that kill mosquito adults or larvae, and the wages of workers who apply those materials on public lands; the reduced weight and milk production in cattle afflicted by these pests—even death from suffocation when mosquito swarms are heavy; drops in tourism; and poor labor efficiency—are all part of the tribute we pay to mosquitoes. That high whine and the silent bite that follows can ruin much more than an evening outdoors. Not only are communicable diseases a threat, but mosquito bites can also send people into shock or lead to secondary infections.

Malaria

Malaria once claimed 100,000 victims a year in the United States. Although it is no longer the danger here that it once was, worldwide it continues to disable up to 300 million people every year. An estimated one out of every three people on Earth has malaria; it causes more deaths, as many as 3 million annually, than any other transmissible disease. Most of those killed by malaria are young children. In adults, malaria is less likely to be fatal, but causes bouts of fever, chills, headache, muscle aches, extreme fatigue, and other symptoms that can occur as often as every two or three days. As a result, in developing countries, malaria is a major factor in hampering economic development and productivity, since it renders millions of people unable to work. Between attacks, victims may suffer extreme fatigue. Stricken families may be able to work only 40 percent of their land, whose output is badly needed for them both to feed themselves and to sell for export. At times, one-third of a developing country's hospital beds may be filled with people suffering from malaria. There is not yet any permanent cure.

The mosquitoes that carry malaria-causing parasites are found in many parts of the United States, but massive public drainage projects and other successful controls of breeding sites have made the disease relatively rare among Americans today. However, during World War II, malaria staged a comeback when over half a million GIs brought the disease home from overseas. During the Vietnam War, as many soldiers were sent home in one year because of malaria as were returned wounded.

Today, most cases of malaria in the United States are transmitted by the needles of heroin addicts. Pockets of it also fester among Asian refugees. Should these unfortunate people come into contact with the mosquito that hosts the malaria organism, the possibility of an upsurge in the infection would be grave.

For centuries, various forms of quinine (a compound originally extracted from the bark of *Cinchona* trees) have been used to ease the chills and fever of malaria. Recently, however, mosquitoes that transmit the disease have been becoming resistant to the pesticide that health workers have been using to hold down their numbers, and the malaria organism that the mosquito injects into human blood is showing increasing tolerance for the drugs doctors depend on to control the disease. Resistance can show up in as few as five years after a drug is introduced.

In the 1970s, experts believed that the use of the insecticide DDT had just about wiped out malaria as a world health threat. Simultaneously, there were increasing concerns about the effects of DDT on the environment and wildlife—especially on birds. Widespread spraying of the pesticide was halted at that time. Sadly, there were enough malaria-infected insects left to start the unholy cycle all over again. Today, malaria is the worst health threat it has ever been in recorded history. Although many earnest environmentalists deplore the use of DDT in developing countries, in these areas it remains the only effective weapon against this deadly menace.

Yellow Fever

Yellow fever, which raged in the United States as late as 1905, is also transmitted by mosquitoes. Fortunately, modern water management has practically wiped out yellow fever in the United States.

Dengue Fever

Widespread in warm climates and carried by the same species of mosquito that transmits yellow fever, dengue is another serious illness. One form, hemorrhagic dengue, causes severe internal bleeding, often leading to death. Unfortunately, other than good nursing care, there is no treatment for dengue. Also known as "breakbone fever," it is a major health problem in Puerto Rico. In April 2001, the U.S. Centers for Disease Control and Prevention established a new facility in San Juan, Puerto Rico, dedicated to the prevention and control of dengue and dengue hemorrhagic fever. Designated by the World Health Organization (WHO) as a Reference Center for Dengue and Dengue Hemorrhagic Fever Research, the facility was given the task of analyzing dengue viruses sent to it from other countries.

Encephalitis

In the United States, the greatest threat to humans from mosquitoes is eastern equine encephalitis, an inflammation of the brain caused by a virus transmitted by a mosquito bite. This disease can be as mild as a slight summer cold—or potentially fatal. Children who contract it, if they survive, are often left with seriously damaged nervous systems. There is no medicine that can prevent or cure eastern equine encephalitis (horses can get encephalitis, too, but they can be immunized against it). The only preventive for us is mosquito control—never-ending, relentless, thorough mosquito control. Encephalitis outbreaks occur every year in the United States, often after a flood. In 1975, one of the worst years, more than 2,000 cases were reported.

West Nile Virus

In 1999, this mosquito-borne disease, a form of encephalitis that is common in Egypt, was recognized in the Western Hemisphere for the first time when sixty-two cases, with seven deaths, were reported in New York City's Borough of Queens. It has quickly surpassed eastern equine encephalitis as the prime form of brain fever endangering Americans. Severe symptoms are thought to be the tip of the iceberg, since many more people with West Nile virus have only mild symptoms or none at all. By 2002, aided by the movement of infected birds, the disease had spread westward. By September 2002, West Nile virus was showing up in nearly every state east of the Mississippi River. Cases have been reported in California, but most cases are clustered in Texas, Louisiana, and Mississippi. In the summer of 2002 alone, forty-eight humans and countless mammals and birds fell victim to this infection. West Nile virus has now reached the West Coast, and the first death in California was confirmed in July 2004. According to the CDC, states with the highest incidence of the disease currently include Colorado, the Dakotas, Illinois, and Minnesota.

West Nile virus is known to be deadly to birds; in fact, autopsies of birds have been a valuable way to track the spread of the disease. If you should find a dead bird on your property, use a shovel or plastic bag to place the bird in a trash bag for disposal. Then alert your state or local health department about your find, and they will tell you how to follow through with a report before disposing of the remains. There is no known danger of humans contracting West Nile virus by handling a dead bird.

CONTROLLING MOSQUITOES

For a while, pesticides brought disease-carrying mosquitoes pretty much under control. When DDT came on the scene, right after World War II, malaria, yellow fever, and encephalitis seemed to fade away rapidly into ghostly threats of the past. As early as 1952, however, a major carrier of encephalitis was showing immunity to the pesticide, and health authorities were worried until new chemicals were developed. Throughout the United States and in other parts of the world, mosquitoes now bounce back from what were once our most dependable chemical weapons. By 1976, California mosquito larvae (the easiest stage to kill since they are not dispersed like the adults) were resisting every available larvicide that does not pose a serious threat to other living creatures.

So today we are back to where mosquito control started a century ago. Our experience with the Panama Canal, when Colonel William Crawford Gorgas of the U.S. Army Medical Corps ordered the isthmus swamps drained and its grasses cut to destroy mosquito breeding sites, was a major breakthrough in the whole area of pest control. It taught us that water management is the key to mosquito suppression. This is as true today as it was then.

The most important aspect of mosquito control is source reduction, which means preventing the larvae from emerging as mosquitoes before they become a threat. Most materials currently used to conrol mosquitoes are biological rather than chemical, so there is little danger to other creatures from toxic fallout. However, as pressure mounts from concerned citizens to preserve small bodies of water as natural wildlife habitats, we may find ourselves coping with the mosquitoes these places harbor. As with all other insect pests, there are no simple answers.

While public health authorities have the responsibility to prevent mosquito-breeding on public lands, you can do a great deal to help.

At Home

All mosquito-control measures boil down to one question: How can we defend ourselves from mosquitoes that are out for blood? Despite massive government efforts at controlling these dangerous pests, they are always with us. And, as we have seen, the available chemical pesticides are losing their effectiveness. What can a person do?

In most areas of the country, the mosquitoes feeding on you are most likely

Mosquito Disease Emergencies

What if encephalitis or dengue fever or some other mosquito-borne disease breaks out in your area, and there is no public agency responsible for mosquito control? What can you do as a citizen?

Turn to your state, which is duty bound to protect its people's health. Request that entomologists and other public-health specialists be sent to your community to monitor and suppress the outbreak. If necessary, your state officials can call on the CDC in Atlanta, which will send in experts to launch the needed program.

breeding in artificial containers or other small sources right near your home, possibly on your own property. To launch your own mosquito-control program, make a stagnant-water survey around your home and eliminate anything or any place that collects water. Correcting these situations is usually a simple matter.

❑ Put empty jars and tin cans out for trash collection.

❑ Discard old tires. If your community does not pick up discarded tires or the local landfill refuses to take them, keep them covered and dry. Use a tarpaulin if you have no covered shelter to put them in. Worn-out water-filled tires are major mosquito breeding sites throughout the United States . . . indeed, the entire industrialized world. Check to see if a local auto-parts dealer will take them from you for recycling.

❑ If there are any tire ruts on your property, fill them with dry dirt or pave over the area, making sure that it has good drainage. Do the same with any puddles under the house.

❑ Get rid of abandoned toys such as wagons, trucks, doll carriages, sand pails, and other items that have been sitting outdoors. Your town may have a toy loan library that will refurbish these and lend them to youngsters.

❑ Empty your children's wading pool at least every three days in summer, and keep it in a dry place the rest of the year. Just turning it over won't do much good since water easily accumulates in sagging vinyl or plastic during a rain.

❑ Screen any cisterns or rain barrels.

❑ If you have a bird bath, empty and refill it at least every three days.

❑ If you keep pets outdoors, change their drinking water daily.

❑ Store wheelbarrows on end, and turn empty plant tubs and watering cans upside-down. If the tubs have drainage holes in the bottom, leave them on their sides so that water doesn't seep up through the holes.

❑ Root outside plants in sand, dirt, or vermiculite. Many people root plant cuttings in water. This is fine indoors, but outside, the water and vegetation are just what an egg-carrying mosquito looks for.

❑ Drain water from tree holes. Fill the holes with sand, cavity cement, or tree sealer (available at nurseries and garden centers). Or drill a small channel from the bark to the bottom of the hole so that water drains away.

❑ If you use plastic sheeting to control weeds, be sure to drain it at least every three days.

❑ Monitor any pools on your property. Although a delight to the eye, an ornamental pool can be a mosquito breeder for the whole neighborhood. Don't despair. You can still enjoy your water lilies. Just check with your local mosquito abatement district to see of they supply mosquito fish to householders. These fish, a kind of minnow, have a voracious apetite for mosquito larvae. Goldfish also eat mosquito larvae, but not as hungrily as the lively gambusia minnow. Another possible source for these fish may be a nearby arboretum or your state's fish and wildlife department.

A properly filtered and cleaned swimming pool will not encourage mosquitoes, but a poorly drained pool deck will. If water regularly collects around your pool, you may need to have the grading and paving corrected. If you have to drain your swimming pool or spa, make sure that no stagnant water remains at the bottom. A partially filled neglected pool is often a source of mosquitoes for an entire neighborhood.

❏ Drain and overturn any small boats or canoes on your property.

❏ Clean dead leaves and other litter from roof gutters at least once a year, and if they need repair, have a sheet-metal contractor fix them as well as your downspouts. Construction faults such as sagging gutters or a flat, undrained roof can attract mosquitoes. Although fixing these problems is usually a small job, you may need to hire a general contractor to bore a hole in a flat roof for better drainage.

❏ Other possible mosquito breeding grounds around a house include fountains, leaky outside faucets, forgotten flower containers on garden workbenches, and the vaults of water meters. A leaky faucet may need only a new washer. Putting a couple of inches of gravel into the water meter vault will make it useless to a pregnant mosquito. As a bonus, this will also discourage cockroaches.

❏ Examine all door and window screens carefully. They are your chief line of defense against flying pests. They should be in good repair, with no tears in the mesh, and they should fit snugly in the door or window frames. When buying screens, make sure that the gauge of the mesh is no coarser than 18 by 16 (the figures refer to the number of individual wires per inch; an experienced dealer will understand the term). Also, screen doors should open outward to prevent any mosquito from pursuing you into the house, and should close completely within four or five seconds. Make certain, too, that all windows and outside doors fit tightly.

In the Community

No matter where you live, nearby public or industrial facilities may be breeding more mosquitoes than any water you or your neighbors are neglecting. Items and places that deserve particular attention include the following:

❑ Storm drains. Scientists at the University of California, Riverside, suspect that most of the mosquitoes in Orange County (a major metropolitan area in southern California) are "undergrounders" bred in the thousands of miles of storm drains and catch basins crosscrossing southern California. The same is probably true of most urban and suburban areas. If your area is plagued with mosquitoes, alert your city or county environmental health authorities. Conscientious health officials routinely treat storm drains with a bacterial larvicide and an insect growth regulator that keep insect young from becoming reproductive adults. They may also introduce guppies and mosquito fish to consume the larvae.

❑ Gutters and potholes. Street gutters that retain water runoff are another mosquito source, as are polluted ground pools, cesspools, open septic tanks, and effluent drains from sewage disposal. Chewed-up roadways, pocked with potholes and cracks, can nurture hordes of mosquitoes. A pint of flood water can support 500 lively wigglers (as mosquito larvae are called) and many a pothole holds considerably more than a pint. If roadways and street gutters in your part of town need repair, and city leaders are sitting on their hands, round up your neighbors to join you in a well-publicized complaint.

❑ Cemeteries. Honoring the dead can be dangerous for the living. In 2002, birds, mosquitoes, and horses in 100 of Illinois' 102 counties tested positive for West Nile virus; the first human cases and deaths were reported in August of that year. By the end of 2002, Illinois had the grim distinction of leading the nation in West Nile virus, with more than 800 human cases and 64 deaths. After surveys showed that many of the victims had recently visited cemeteries, state authorities launched a campaign to eliminate mosquito breeding sites on cemetery grounds. Among other changes, mourners were asked to stay off the grounds at dusk and dawn, when mosquitoes are apt to be active. Another recommendation was that managers encourage plastic flowers, potted plants, or small flags as grave decorations, as opposed to cut flowers in water-filled vases.

❑ Technology's throwaways.

• Tires. The industrial world's advanced technology and quick replacement of worn-out parts are sabotaging our ongoing war against mosquitoes. The most serious problem is worn-out tires. One water-filled castoff in La Crosse, Wisconsin, for example, was found to nurture 5,000 mosquito larvae, a dangerous number of which carried viruses that could cause encephalitis. The Asian tiger mosquito, an especially dangerous species, has a particular fondness for the water in discarded tires, but other species also are drawn to tires. Some 240 million tires are discarded in the United States every year. Because tires tend to rise to the top of a sanitary landfill, many city dumps don't accept them, so they are just tossed onto vacant land. There they collect water, ready for any egg-carrying, disease-carrying mosquito. All-rubber tires are easily recycled, but steel-belted radials resist recycling in a cost-effective form. Environmental specialists are working on this problem and turning up some ingenious solutions. For instance, if chained together, hundreds of these tires form an effective barrier reef for an oceanside community. Or if weighted, chained, and dropped into a freshwater lake, the tires make a fine spawning ground for catfish and bass. They can also be incorporated into asphalt to pave roads. The need to conserve petroleum may ultimately help us solve our discarded tire problem completely. In the 1980s, a New York–based firm began burning tires from a huge tire dump near Modesto, California, to produce electricity, which it sells to the local utility company. The company met state air-quality standards. In 2002, cities on both sides of the United States and Mexican border were looking into feasible ways to recycle tires.

• Abandoned automobiles, refrigerators, and buildings. Derelict cars and discarded refrigerators, especially those with plenty of organic matter inside, can produce swarms of mosquitoes. Deserted buildings are another fertile source; containers gathering even a little bit of still water outside abandoned structures, dilapidated roofs, and roof gutters will also attract pests.

• Industrial plants. Living within a mile or two of a factory can aggravate your mosquito problem. Industries often use great volumes of water, and also maintain sumps, pits, and water towers. Improperly drained industrial waste water can launch squadrons of mosquitoes to nearby homes.

Whether the nuisance is a house, a store, a refrigerator, a car, or a carelessly run factory, protect your family and yourself with a call to the public health authorities.

Special Concerns for Rural (or Recently Rural) Areas

As residential areas spring up farther and farther out from the cities, mosquitoes become more of a pest. In rural or semirural environments, suppressing them is more complicated than it is in urban or suburban settings. Besides having to watch all the trappings of family living, farmers and other rural householders should be aware of the many agricultural processes that foster mosquitoes.

Dairies

Dairies and other cattle facilities use huge amounts of water to wash their animals, cows, barns, and equipment. The resulting water is rich in organic matter that nourishes mosquito larvae. Often, the water is dumped on impenetrable ground, where it forms stagnant pools. If a nearby dairy is sending clouds of mosquitoes—

or even just a few aggressive females—your way, let your local mosquito abatement district know about it. They can have the situation corrected.

Irrigation Facilities

As we have moved to drier, warmer climates, we have developed vast irrigation networks to turn these former deserts into bountiful farms. If water doesn't drain properly but stands in ditches, laterals, and canals, or gathers in pools below the watered fields, ten irrigations a year can produce ten generations of mosquitoes in the same twelve months. As entomologist Walter Ebeling observed, "Swarms can become so dense that one's clothing becomes covered with a continuous layer of insects, and livestock may be killed by their continuous attacks."

Ponds, Watering Holes, and Hoofprints

There can be other mosquito sources in rural areas, such as hoofprints around watering troughs and livestock watering holes that have plants clustered around their banks. Drinking troughs should be set on gravel or cement to prevent pooling. Any ponds should be deep and straight-sided to discourage the growth of plants in which mosquito larvae can hide from predators. A population of greedy mosquito fish also will keep a pond clear of wigglers. If you run a farm or dairy (or live near one) that is having a mosquito problem, your county agricultural agent or mosquito abatement district can help you eliminate the pests.

Nature's Controllers

Nature has some potent mosquito controllers that have been on the job for millions of years.

Birds, Reptiles, and Predatory Insects

Purple martins, whippoorwills, and hawks all eat mosquitoes. In addition, mites, ants, and booklice relish mosquito eggs, while frogs, turtles, and a large, harmless species of mosquito prey on the dangerous species' larvae and pupae. Spiders, mites, geckoes and other lizards, and bats also devour adult mosquitoes.

Dragonflies make short work of both larvae and adults. A number of towns in Maine buy dragonflies by the thousands, following the lead of the town of Wells, Maine, which has used these insects for years as a major means of mosquito control. These insect helicopters can stop in midair and change direction in an instant; their speed has been clocked at 35 miles an hour.

The problem with all these predators is that they may not eat the mosquitoes we humans want eaten. We can't confine the birds, bats, dragonflies, and lizards to the areas we think need cleaning up.

Mosquito Fish

Gambusia minnows, or mosquito fish, are used extensively in ponds to keep down mosquito larvae, for which they have a voracious appetite. This is the one predator that we can place exactly where it will do us the most good. Many mosquito abatement districts supply these fish to citizens free of charge. Another possible source for the fish may be a nearby arboretum or your state's fish and wildlife department. The city of Colorado Springs, Colorado, bases much of its mosquito control program on this minnow.

PERSONAL PROTECTION

Common sense tells us to stay indoors when mosquitoes are most active. In most areas, they come out to feed at dawn and dusk, and stay active during the night. If you must go outdoors during the day in a place with a known mosquito problem, it is wise to protect your skin with an effective repellent.

Chemical Repellents

There are a variety of repellents that have been proven effective in deterring mosquitoes. Those with proven track records are described in the following sections.

DEET

The most effective mosquito repellent, sold as a spray, stick, or lotion, is , N,N-diethyl-meta-toluamide—popularly known as DEET. DEET was first used by American troops in Vietnam, and it is the standard against which the United States Department of Agriculture (USDA) measures all newly developed repellents.

DEET will keep mosquitoes at bay for two to twelve hours. It is sold as DEET and is also the active ingredient in Off and other brand-name insect repellents. In these commercial products, DEET is often mixed with substances that repel insects other than mosquitoes, thus giving good all-around protection.

According to toxicologist Marc Bayer of the University of California, Los Angeles School of Medicine, DEET should not be used by chronically ill people or on very young children. Further, it should not be applied for more than a few days at a time.

The National Pesticide Information Center (NPIC) states that DEET should be applied to exposed skin only. DEET should *never* be applied to skin that will be covered by clothing. Doing so has led to at least one case of nervous system damage. In addition, it should not be sprayed on shoes or clothing, as it can last indefinitely.

Because DEET will not harm nylon, it can be applied over 100-percent nylon hosiery. However, it may break down lycra fibers, and it can damage plastics, such as those used to make watch crystals and pens, as well as rayon fabrics. It also dissolves paint, varnish, and leather. On a trip through the mid-Atlantic states, I learned not to carry DEET unprotected in a leather handbag. A leaky applicator ruined my purse.

Whenever using any product containing DEET, be sure to follow the label directions carefully.

Two other names of chemical repellents that you may find on product ingredient lists are dimethylphthalate (DMP) and ethohexadiol (also known as avermectin B1 and Rutgers 612). These have largely been supplanted by the more effective DEET, however.

Permethrin

Permethrin is a recent addition to our arsenal of chemical repellents, supplementing the protection of DEET. Permethrin was originally derived from the chrysanthemum plant, and can be safely used even around most children (although some people may be allergic to it). A baby can suck on netting treated with permethrin and not get sick. Whereas DEET works best when applied to bare skin, permethrin is applied to fabrics. Since mosquitoes can bite through thin materials, treating clothing with permethrin and bare skin with DEET gives excellent protection against insects.

Various clothing retailers are now marketing garments impregnated with permethrin as effective insect protection. These garments are not for long-term wear; tests at the USDA's Center for Medical, Agricultural, and Veterinary Entomology (CMAVE) in Gainesville, Florida, have shown that laundering greatly reduces the effectiveness of such clothing.

When applying a chemical repellent—any repellent—on a child, avoid getting any of it on the youngster's hands. Any repellent can irritate sensitive tissues, and children often pop their hands into their mouths or rub them over their eyes.

Because mosquitoes are potentially so dangerous to humans, you must weigh the risks of getting a lifelong disabling disease against the risks of accidentally misusing an effective repellent. Careful use of a repellent should enable you to avoid both potential sources of danger.

Special Concerns When Traveling Abroad

In many countries, especially less developed ones located in warmer climates, door and window screens are a luxury beyond the means of most people. If you should visit such a place, be aware of the lack of screens and the subsequent greater risk of mosquito-borne disease. Dengue fever, for example, occurs fairly frequently in the West Indies and the South Pacific.

Before going to a tropical country, check with your doctor or public health department about immunization against illnesses that could ruin your holiday— and stay with you for life. A dermatologist friend of mine contracted encephalitis on a holiday in Costa Rica and has been disabled ever since. For some mosquito-

Starting a Mosquito Abatement District

By now, you have gathered that mosquitoes are very much a public health concern, and controlling them is a public responsibility. Many towns and cities have established agencies to keep these insects within tolerable bounds. If you live in a small community that has no central mosquito control agency, your best long-range solution is to start a mosquito abatement district. Every one of the fifty United States has legislation permitting localities to establish abatement districts.

To form such a body, citizens petition their city council or county commissioners to set up a special tax for funding a mosquito control agency. The bureau is staffed by a manager, a source-reduction expert (someone who can spot a mosquito breeding site and knows how to dry it up), an entomologist (insect expert), and inspectors and supervisors to see that the necessary drainage or chemical applications are carried out.

Forming such a district takes time, but after you have submitted your petition, you don't have to just sit idly by and endure mosquitoes. Many organizations have a real interest in maintaining your community as a comfortable, healthful place in which to live and do business. As a local resident and taxpayer, you can call on the local chamber of commerce, realty board, planning commission, school board, and street and road department, as well as civic groups and city and county health departments, just to name a few.

If you live in a rural area, farm bureaus, dairy leagues, cattle-herders' associations, and the Grange all have a vital stake in suppressing mosquitoes. Urge their members to help you publicize the problem—and its solution—through newspaper stories, radio and television programs, and school classes or meetings. Figure 7.2 is a leaflet formerly used by the U.S. Centers for Disease Control and Prevention (CDC) to alert communities to the need for mosquito control and to tell them how to implement such controls. You may copy and use it if you wish.

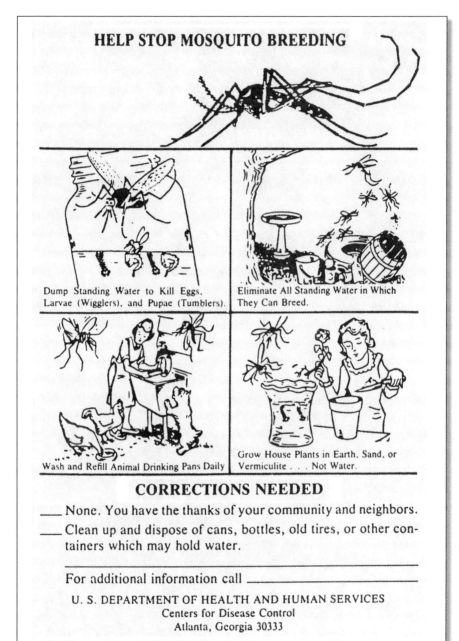

Figure 7.2 An Educational Leaflet Formerly Used by the National Centers for Disease Control

Source: United States Department of Health and Human Services (Atlanta: Centers for Disease Control) leaflet.

transmitted diseases, there are no vaccines and no specific medicines. Your best—indeed your only—protection is a good repellent.

If you are planning a visit to the tropics, you would be wise to include a permethrin-impregnated mosquito net in your luggage. The mesh should be no larger than 1.2 by 1.2 millimeters. A well-stocked sporting goods store should carry mosquito nets; you can also buy them through the Internet. You could apply the permethrin to the netting yourself, but it is certainly more convenient to buy one that is already treated.

WORTHLESS "CONTROLLERS"

Many commercial products claiming to repel mosquitoes come and go. According to the Department of Entomology at the University of California, Davis, most of these are worthless.

Herbal Products

Whether they are wrist bands impregnated with aromatic oils, vitamin B_1, or mixtures of brewer's yeast and garlic, they have little to no proven effect on mosquitoes. Oil of citronella, burned in candles and long touted as an effective repellent, should be used only out of doors—never indoors—and even then, only if there is no breeze. Whether in the form of a candle or an electric coil, citronella works only if the air is still. It seems hardly worth the trouble and expense.

Electrical "Controllers"

A number of electrical devices claim to control mosquitoes. They range from the pretty useless to the utterly useless.

Electromagnetic Repellers

Various electromagnetic insect repellers have been tested and found worthless. When researchers tried one device that claimed to generate sound waves mosquitoes don't like, nearly 89 percent of the confined mosquitoes took blood whether the machine was turned off or on.

Insect Electrocutors

Called Bug Blaster, Bug Whacker, Zapper, and the like, insect electrocutors do kill flying insects. They attract the creatures with ultraviolet light and then electrocute them. These products, however, are not much help against mosquitoes.

In tests conducted in 1982 by scientists at the University of Notre Dame, only 3.3 percent of the insects killed by these devices on an average night were female mosquitoes. Human beings near the electric insecticides proved to be more attractive than the ultraviolet light, and people had just as many mosquito bites after the appliances were installed as they did before. The devices also kill valuable predator insects, and may actually attract more insects to your yard than you would have otherwise.

POSSIBLE FUTURE CONTROLS AND REPELLENTS

Scientists have been investigating other natural controls for this eternal menace—controls that pose no danger to other life forms.

Biological Controls

One of the most promising biological controls is a bacillus (a type of bacteria) developed in Israel that infects mosquito young with a fatal disease. Another substance that blocks the maturing of larvae is also available. Both of these, however, are best suited to wide applications, as in large accumulations of water. A third mosquito-control technique under study is the release of sterile males.

Certain algae and freshwater fungi have been found to be poisonous to immature mosquitoes. The gluey seeds of a common California weed have been tested for their ability, when placed in water, to attract mosquito wigglers, which then stick to the seed and die from drowning or exhaustion. One pound of the seeds is estimated to wipe out 54 million larvae.

Catnip

This is one of the most promising developments in mosquito control. In 2002, researchers from the Department of Entomology at Iowa State University announced that nepetalactone, the active ingredient in catnip, is about ten times more effective than DEET in repelling mosquitoes. It is believed to be nontoxic to humans (and is also effective in repelling cockroaches). It is not yet on the the market, but the compound has been patented and we can hope that it will gain FDA approval and become available to the general republic.

Distraction

Researchers at the Center for Medical, Agricultural and Veterinary Entomology are following a novel approach. They are trying to develop a chemical formula that will be so distracting to mosquitoes that they will forget about their human prey and try to find the source of the smell. Chemist Ulrich Bernier is pinpointing a few chemical compounds given off by our bodies that mosquitoes use to find us; his research aims at making attractants that are nontoxic to us but confusing to the mosquito, and putting together a mixture that is an effective, nontoxic repellent.

These approaches may have a significant role to play in the future. Today, however, water management is still the sure preventive. Monitoring the number and kinds of mosquitoes in an area; the strategic use of predator fish, screens, and repellents; and expert use of selective pesticides by trained professionals also are used today to good effect. But, despite all our science and technology, the mosquito

What If Mosquitoes Find You Especially Attractive?

Why is it that in a circle of friends chatting outside, one or two may be heavily bitten while the rest sit comfortably, apparently immune? Scientists tell us there are malaria and yellow-fever carriers who would rather feed on a chicken, a cow, a pig, or a horse than a human but, lacking the livestock, will settle for a man, woman, or child.

As far as researchers can determine, the combination of warmth and moisture is very alluring to mosquitoes. So is carbon dioxide. Dry ice, a form of carbon dioxide gas, is used to trap the insects when scientists want to know what species are in a given area. Since we give off carbon dioxide with every breath, all of us are potential mosquito meat.

However, other factors enter the picture. Some species like light-colored clothing, some dark. Yellow- and dengue-fever mosquitoes tend to prefer men to women, and an ovulating woman is more likely to be bitten than one who is menstruating. This seems to be tied to the amount of estrogen in the blood. All in all, warmth and carbon dioxide, enhanced by amino acids and estrogen, appear to determine our appeal for mosquitoes.

If you are someone mosquitoes find delicious, here are some pointers that may help you avoid being bitten:

- Stay indoors when mosquitoes are active. (You may need to get in touch with your local health department to find out when that is.)

- Use a proven repellent (see page 125).

- When outside, wear long-sleeved clothing made of a tightly woven fabric. Keep sleeves and collars buttoned, and trousers tucked into socks. This is the method used by public-health workers in areas known to harbor disease-bearing insects.

remains a major threat to human health and comfort. Mosquitoes abound in such astronomical numbers that we will probably always be warring with them.

To keep the fight even, we must be ever alert to the ways in which we handle water, from the smallest outside puddles to the vast salt marshes that rim our coastlines. Strict water management will enable us to coexist safely with this graceful, persistent, and potentially deadly enemy.

8. The Mighty Flea, the Insidious Tick

The great Russian composer Modest Moussorgsky found the flea a most amusing creature and wrote a song for baritone, "The Song of the Flea." It's a funny song, but there is nothing funny about fleas. Many people who love cats and dogs are reluctant to adopt one because they don't want to cope with the fleas and ticks—and the chemical pesticides—that are often part of a pet's life. Keeping pets free from these bloodsucking parasites without resorting to insecticides does take some time and work. Although millions of dollars' worth of pesticides are sold each year to control fleas, pet owners still groan about their flea problems. Ticks don't get as much attention, but they can be as dangerous and bothersome as fleas. Fortunately, though, some safe procedures, if followed regularly, can control both of these pests effectively—and *without* the use of toxic chemicals.

FLEAS

The seventeenth-century Dutch artist who sketched the first flea seen under a microscope was heard to mutter while peering through the lens, "Dear God, what wonders there are in so small a creature!" He spoke the truth.

Looking rather like a flattened armor-plated tank, these wingless insects have strong "thorns" and hooks on their heads and thoraxes that hold them fast to their furry hosts. (See Figure 8.1.) Although an individual flea is barely $\frac{1}{64}$ inch tall, its powerful legs can shoot it seven or eight inches up into the air and fourteen inches sideways. By way of comparison, a human with such strength could clear the top of the Washington Monument or Egypt's Great Pyramid in one bound.

A flea can pull 400 times its own weight and lift an object 150 times heavier than itself. By contrast, a horse can pull four times its weight, and we humans have difficulty lifting an object equal to our body weight. In addition to having great physical strength, fleas can (according to one authority) be frozen for hours or days and thaw out to be as good as new. Moreover, one drop of a flea's saliva is enough to start 100,000 people furiously scratching.

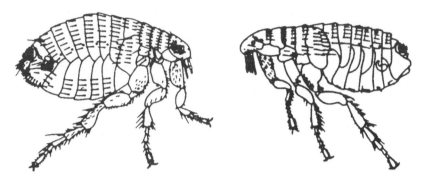

Figure 8.1 Two Types of Fleas

Source: Fleas of Public Health Importance and Their Control (Atlanta: Centers for Disease Control, Homestudy Course 3013-G, Manual 7-A, 1982).

Types of Fleas

Of the estimated 1,600 known kinds of fleas, only a few types are important for the average person.

Cat Fleas and Dog Fleas

Cat fleas and dog fleas are similar in appearance and, despite the names, both species may be present on either dogs or cats. They can infest many other animals as well, including chickens, rats, opossums, racoons, and foxes. They readily bite people, and are the types of fleas most likely to bother you. The larvae look like tiny white worms, and they prefer a moist environment of dust and organic debris. They are very common in yards and under houses. However, they can develop only in places where there is at least 75 percent relative humidity, so don't expect to find them in the hot, dry portions of your yard.

Human Fleas

At one time, the human flea was the type that most commonly annoyed people. Its bite can cause a severe allergic reaction. Upper-class Eureopan women in the Middle Ages often passed the time searching themselves for fleas, and would bestow on favored suitors the privilege of helping them in the hunt. Some historians speculate that lap dogs became fashionable as unobtrusive flea traps. Because of their exceptionally strong legs—even for a flea—human fleas were the star of the flea circus, glued for life to a tiny chair or tethered to a lilliputian coach or cannon.

Homemakers used to damp-wipe their wooden floors and furniture, providing fleas with the humidity essential for their reproduction. Since the advent of the vacuum cleaner, damp-wiping is no longer so widely practiced, and this species of flea is now found mostly on pigs.

Chigoes

If you live in the tropics, you may have the bad luck to meet up with a chigoe (or chigger), a nasty flea that burrows between people's toes and under their toenails, causing intense pain and, ultimately, even gangrene. It is the inspiration for the expression, "I'll be jiggered!" The best prevention for a chigoe bite is to avoid walking barefoot and to wear closed shoes if you are in an area known to have chigoes. When you are on vacation, ask some of the locals if you need to watch out for chigoes.

Sand Fleas

The term "sand flea" actually refers to a special species. In the northern United States, they are cat or dog fleas found in vacant lots, where they have dropped off of stray animals. In the West, they are cat or human fleas that have fallen from deer, ground squirrels, or prairie dogs. In the South, they may be poultry fleas (also called "sticktights") or cat or dog fleas. On the beaches, sand fleas are harmless crustacea, rather like tiny shrimp.

Snow Fleas

A snow flea is not a flea at all, but a springtail—a tiny soft-bodied, wingless insect. By suddenly releasing a long appendage tucked under its abdomen, this animal can spring into the air, causing it to be mistaken for a flea.

In addition to the creatures described above, rat fleas, chief carriers of bubonic plague, are found wherever there are rodents. Lately, with a rise in the number of plague cases, we have had to pay more attention to rat fleas.

Although each species of flea prefers to feed on the animal from which it takes its name, all of them can shift very quickly to another mammal when its host dies.

How Fleas Make Themselves at Home

Female fleas lay their eggs either on or off a host animal. Those laid on a host animal roll off onto the animal's bedding or nest, or accumulate in carpets, floor cracks, furniture, and dust, and—outside—in damp soil, manure, and other organic debris.

Depending on the area's temperature and humidity, the larvae hatch anywhere from two to fourteen days later. They prefer temperatures between 66°F and 84°F, and humidity between 70 and 90 percent. Flea eggs never hatch when the thermometer falls below 40°F, and most larvae die when the humidity is less than 40 percent. This is why a warm and wet winter or spring usually foreshadows a bad flea season.

"In warm weather the average dog has sixty fleas, half of them female," says Dr. Steve Wagner, former president of the California Veterinary Association. "A flea can lay 600 eggs a day. That's 18,000 eggs a month."

The eggs usually hatch in carpet dirt, under beds, and in basements. The young of some species feed on the dried blood excreted by the adults. The "pepper and salt" often found where an infested animal sleeps is actually a mix of flea eggs and this dried blood.

Anywhere from nine to two hundred days after hatching, the larva spins a cocoon and becomes a pupa. This pupal stage can last from one week to one year, with the adult flea often resting inside the cocoon for many months, ready to pop out as soon as it senses vibrations, warmth, or carbon dioxide from a nearing host. The ability of this insect to prolong its life span is at the root of the flea explosions that occur every summer, when warmth and humidity are just right for flea survival.

This is the reason why many people with cats or dogs, on returning from a vacation during which their pet was boarded, are attacked by hordes of hungry fleas. And the fleas don't care who owns the pets.

A Florida man was plant-sitting for pet-owning neighbors on a holiday and, while watering the house plants, felt a stinging sensation over his legs. Looking down, he saw that his calves and ankles were covered with feeding fleas. It's a universal experience. Householders in ancient Egypt reportedly used to smear a slave with donkey's milk and stand the luckless soul in a corner of the room as a living flea trap. In Arab countries, it once was the custom to pay an unemployed man to enter every room of an uninhabited house to lure all the fleas from their hiding places. Canny country folk in the southern United States will set a sheep in an empty building as flea bait.

The Hazards of Fleas

Although it has fallen behind the mosquito as the world's most dangerous insect, (currently, mosquito-caused deaths total millions annually; those from fleas, thousands), the flea is still a catastrophe waiting to happen. The plague that fleas can carry is an infection that can be transmitted from sick mammalian hosts like rats or ground squirrels to human beings. Throughout history, plague has periodically depopulated whole nations, even continents. In the sixth century, plague killed almost half the people in the Roman Empire. Eight hundred years later, probably helped along by the Crusaders traveling between Europe and the Middle East, an estimated three-fourths of the people in Europe and Asia succumbed to it.

Although plague attacks and kills many fewer people today, it is still a danger, more so in developing countries than in industrialized nations. However, the western half of the United States is still one of the world's major reservoirs of this fearsome disease. In just two months—from April to June 1983—there were sixteen cases in Arizona, New Mexico, Utah, and Oregon. During the 1980s, plague cases in the United States averaged about eighteen a year, most of them people under twenty years of age. Hunters who skinned squirrels and other rodents were the most likely victims. About one in seven persons who contracted the disease died. Plague-infected squirrels have been found within the Los Angeles city limits, and there have been human cases, transmitted by cats and dogs, in nearby suburbs and mountains. Whole sections of the city have been quarantined and declared off-limits because of plague.

The infection can take one of three forms: bubonic, pneumonic, or septicemic. Since the development of antibiotics, bubonic plague (the dreaded Black Death) is much less deadly than it once was. If untreated, however, this disease has a death rate of 25 to 50 percent. Pneumonic and septicemic plague are almost always fatal.

Once a scourge encountered in seaports with their ship-infesting rats and the fleas they carry, plague is now primarily a rural problem, carried by the parasites of chipmunks, squirrels, opossums, and other small mammals. By killing and/or eating a plague-infested rodent, a cat or dog living in or near a wilderness area can readily pick up infected fleas and bring them to the pet's owners. Such a case happened in Lake Tahoe, California, in 1983, when a young schoolteacher contracted pneumonic plague from her cat and was diagnosed too late for medicines to help her. The disease can be transmitted from one person to another by means of infectious droplets from coughs or sneezes. Fortunately, in the case cited above, authorities were able to protect the teacher's pupils with antibiotics.

If you live in or near a western wilderness area and have pets, keep them confined, and rodent-proof your house as described in Chapter 5. And, of course, don't make friends with any ground squirrels or chipmunks.

In addition to plague, fleas can carry other serious diseases, among them typhus, tularemia, and tapeworm. The bites of these insects can also provoke strong allergic reactions.

Controlling disease-bearing fleas on wild animals is a problem for public health authorities. Controlling fleas on your pet and in your home is your problem, one that can be a real headache. They are among the most stubborn nuisances we encounter.

There are an estimated 110 million domestic dogs and cats in the United States. That comes to about one animal for every two Americans, not counting the feral ones. Not surprisingly, flea control is a big business. Thousands of people make a good living turning out antiflea powders, lotions, shampoos, and collars; dipping infested pets; compounding medicines to heal animals who have ripped their itching skin in frenzy; spraying our homes and yards with chemicals to stem the flea onslaught. At best, they have just modest success. Even treating a whole house gives only short-term relief.

What makes this pest so invulnerable? The secret lies in the insect's life cycle, its ability to survive for long periods without food, and its readiness to move from one kind of host to another.

SHUT THEM OUT

Preventing a flea infestation is probably easier than getting rid of these pests. The simplest way, of course, is not to have a pet or other domestic mammal in your home. Those for whom a loyal dog or affectionate cat is one of life's great pleasures are likely to face a flea problem at some time in their animals' lives unless they establish and hold fast to some ground rules as soon as they bring home their first puppy or kitten. Here they are:

❏ Don't let your dog or cat roam the neighborhood. It can pick up fleas from infested gardens or other animals.

❏ Decide at the outset if your cat will live inside or outside. People whose cats come in and out at will are more likely to have flea problems.

❏ For a dog that spends much of its time outside, provide a fenced exercise area with a sturdy dog house.

❏ In warm weather, comb your pet every day with a flea comb.

❏ Bathe your pet frequently in summer, when fleas are most active, and every two to four weeks in winter. If you adopt a kitten, get it used to regular baths before it becomes too opinionated on the subject.

❏ Vacuum your pet's sleeping area several times a week.

❏ Use towels to cover your pet's sleeping cushion and wash them several times a week.

❏ Clean pet bedding often.

Even if you don't keep a cat or a dog, you may still find yourself coping with fleas. Where in the world can they be coming from? There are several possibilities, among them the following:

• You may have visited a friend with a flea-infested pet.

• Your child may have played with a neighbor's flea-infested dog or cat and carried the insects home.

• An infested cat may have had a litter of kittens in the crawl space under your house or porch. (When the kittens went off on their own, the flea eggs hatched, and now the adult fleas are looking for the nearest blood meal.)

• A squirrel or other rodent may have once nested in your attic and, when it was excluded, its fleas and their young remained behind.

• In rural areas, the source could be an opossum in the wall voids of your home.

• Whether in the city or the country, infested rats in the wall voids or nearby could be the origin.

Eliminating these sources would be an effective first step in getting the situation under control.

WIPE THEM OUT

Until recently, available insecticides could be cruel protection for pets and their owners. The cat or dog being sprayed or powdered with chemical insecticides was at considerable risk of poisoning. The animal's owner may have been, too. Cats and dogs, says Dr. William B. Buck, veterinary toxicologist of the University of Illinois, were frequently poisoned by organic insecticides. The chemicals most commonly used in antiflea collars, dips, sprays, shampoos, and powders contained substances known as organophosphates and carbamates. These substances cause most of the pet poisonings. Daily dusting with a carbaryl compound (a type of carbamate) can kill your pet.

If you sprayed or powdered your pet frequently with an organophosphate or carbamate during flea season, you may have observed that you, your pet, or both of you became noticeably irritable. These chemicals attack the nervous system, a good reason not to use them.

A final word on conventional flea insecticides: *They may not bother the fleas at all, but you may have a severe reaction to them.* One Los Angeles woman sprayed her baseboards to prevent fleas, even though there was no infestation. She used a leaky bottle, and didn't pay much attention to the pesticide spilling onto her skin. Soon after, she developed a painful, itchy eruption on her hands, a rash she still had when I met her two years later.

Some time ago, a delightful cat adopted us. She got all of our attention—that is, when we weren't scratching flea bites. Although I powdered her every week with a carbamate the veterinarian had sold me, the fleas just grew fatter and sassier.

Unwilling to fill the house with the fumes of a flea bomb (a friend had been made sick by one), I bought a highly recommended spray that I was assured was the safest thing around, and treated cracks, rugs, and other flea hideouts. I applied the chemical strictly according to the label's directions. It did the trick. The fleas were completely routed—but our son nearly was, too. He had a strong allergic reaction to the product, with swollen eyes, puffy lips, and difficulty in breathing.

Fortunately, in the 1990s three new products came on the market that have made flea control a lot safer for pets, as well as safer and more convenient for their owners. Their technical names are fipronil (sold under the brand name Frontline), imidacloprid (Advantage), and lufenuron (Program). Advantage protects only against fleas; Frontline provides long-term tick control along with flea control; Program prevents any eggs a female flea lays on your pet from hatching.

Advantage has the virtue of killing nearly all the fleas on an animal within twenty-four hours. A single dose lasts for four weeks on dogs and up to four weeks on cats. However, your pet's skin can become quite greasy for a few days. Also, if you have children who like to cuddle with the animal, they may be sensitive to the pesticide, developing a rash or some other symptom.

Frontline can kill fleas for up to three months on dogs and a month or more on cats. This product also kills ticks for a month or more.

Both Frontline and Advantage are "spot-on" products applied directly to the pet's skin. Program, administered by mouth, produces sterile flea eggs within seven days after your dog or cat has taken it. However, if you already have a flea infestation, it will take as long as two months to reduce the flea population to a minimal level. And if you live in a warm climate, you may need to use Program year round.

There seems to be no data on the effects of human exposure to Advantage or Frontline. Since you never come into direct contact with Program, there is little danger of a toxic reaction. As with all chemical pesticides, we can expect the fleas to develop resistance to them, a major problem in all aspects of pest control.

All of these products must be bought from a veterinarian, as often as every month. Not surprisingly, flea control has become a major profit center for veterinarians, but the cost may be worth the results: convenience and peace of mind. However, if you are being careful with your money, you may prefer to use more economical routes.

Considering the flea's incredible jumping ability, its widespread and varied hiding places, and our animals' fondness for roaming, it is probably impossible to have a pet that is always completely free of fleas. However, to keep your pet and yourself comfortable without exposure to pesticides, you must use three overall strategies: environmental controls (both indoors and out); repellents; and mechanical controls (such as traps, combing, and bathing.)

No single strategy will bring you complete success, and, because animals and people vary greatly in their sensitivity to insect bites, you have to experiment to see which combination works best for you. While searching for that mix, don't get discouraged. According to veterinarian Steve Wagner, at least 40 percent of the commercial products sold over the counter for flea control are ineffective.

Environmental Controls

"If you're talking flea control," says veterinarian George Peavey, "you're talking environmental control." Fleas can be in your upholstery, your carpets, the cracks and crevices of your parquet floor, and your yard. They may also be in your attic, basement, and wall voids. (Remember, even if you don't have any pets, you could have a flea infestation.) Use the checklist that follows for environmental control measures:

❏ Check the immediate area for rodent nests, and destroy any that you find. They may be empty, but fleas can still be hiding in them. If they are nesting under your home, clear them out.

❏ Keep your grass cut short so that the sun can get to flea larvae; at this stage, the pest cannot survive in exposed sunny areas.

❏ Since fleas rest in organic material, use overwatering or drying to kill them in your yard, advises the Center for the Integration of Applied Science in Berkeley.

❏ Make sure that the outside structure of your house is in good repair. Foundation, foundation sills, windows, doors, and vents should all be tightly sealed. Inspect your chimney and the crawl space under the subflooring, and close off any cracks and holes.

❏ Rodent-proof your home as recommended in Chapter 5. Be especially careful if you have an attached garage, a frequent entryway for flea-carrying rats and mice.

❏ Make sure that no stray cat can find a niche under or around your home for giving birth.

❏ If the fleas are coming from a stray animal that once found shelter in your house, you may need to treat your attic and wall voids with a long-lasting silica gel. This material, though nontoxic, is very irritating to the lungs and nasal passages, so wear adequate protective gear when applying it. Or you might hire a pest control operator to do this for you.

❏ In summer, when fleas are at their peak, consider keeping your pet outside all the time. Keep the animal comfortable with frequent baths and combings and by regularly vacuuming its sleeping place and laundering its bedding.

❏ Indoors, your vacuum cleaner and a good flea comb are the most important part of nonchemical flea management. Vacuum any infested rooms of your house frequently, daily if necessary, using the crevice attachment to suck up insect adults, larvae, and eggs from upholstery crevices, carpet corners, loosened baseboards, and floor cracks. Most flea larvae will be found within a foot or two of a pet's resting spot. For carpeting, use a vacuum with a rotating beater bar. Place each day's collection of dust and insects in a tightly closed bag, and leave it to "bake" in a covered garbage bin in hot sun.

❏ To bring a heavy infestation down to tolerable levels quickly, have your carpets and upholstery steam cleaned.

❏ Keep your pets away from the basement and attic, both of which are hard-to-clean areas.

❏ Bathe your pet frequently with a gentle shampoo. (Since fleas drown easily, it is not necessary to use an insecticidal shampoo.)

❏ If your pet sleeps inside, cover its resting places with removable, washable cloths. Wash them every couple of days, since flea eggs can hatch two days after being laid. Or just spin the cloths in your dryer for a half-hour. Alternatively, you can hang them on a clothesline in the sun for two days so that the sun can dry and kill any stages that may remain.

❏ Try sprinkling borax on carpets and leaving it there for twenty-four hours before vaccuming it up. Some pet owners have found that this kills fleas.

❏ If you don't have a dog or cat, but still have fleas in an upholstered piece of furniture, you may have brought the insect in on your shoes. If the infestation is small, running a steam iron lightly over the chair or sofa will clear them out quickly and safely.

In the 1930s, Hugo Hartnack, an early practitioner of integrated pest management and self-styled "economic entomologist," had these suggestions for fighting fleas. (Few of us sleep on straw mattresses today, but his other ideas are still good.)

❏ Have a cement-floored basement.

❏ Don't use straw mattresses.

❏ Have crack-resistant, smooth hardwood floors.

❏ Clean uncarpeted floors with an oil-based product instead of water.

❏ Vacuum all rugs and carpets frequently.

❏ Use central heating to dry the air in your home.

❏ When taking a new pet into your home—or even if you allow a stray to live on your property—be aware that the animal may have fleas. Examine it and treat it as necessary.

❏ Heat your new residence to 122°F for several hours. This, Hartnack's final suggestion, would be useful before you move into a new home. Do it before your furniture arrives.

Natural Repellents

Over the years, people have found several of the repellents discussed below to be effective, at least to some degree. Veterinarians may tell you that they don't work, but they see only animals with serious flea allergies. Cats or dogs without fleas, or those untroubled by their bites, don't need a veterinarian for flea allergies.

Veterinarians are divided on the value of brewer's yeast. Some say it's useless; others think it is well worth a try. Prominent veterinarian and author Dr. Wendell Belfield, while not claiming miracles for brewer's yeast (if you have read the rest of this book, you know that there are no miracles in pest control), says that some cats and dogs respond well to this rich source of the B vitamins. He suggests starting to give it to your pet in spring, and continuing all through the warm weather to build up and maintain your pet's flea resistance. The yeast apparently imparts an odor to the animal's skin that fleas don't like. Dr. Belfield suggests a daily dose of 25 milligrams per ten pounds of the animal's body weight.

A word of caution: If brewer's yeast is given in large doses or with dry food, it swells, causing the animal considerable intestinal discomfort. We mixed one-half teaspoon daily in our cat's moist food for months. She accepted it readily and had no cramps or gas.

There seem to be as many herbs suggested for flea control as there are pet owners. Everyone has a favorite. Some herbs are mentioned more frequently than others, so these probably ought to be the first you try. Many people claim that pennyroyal mint (which can cause a rash), eucalyptus, citronella, and rosemary sewn into a collar worn by the animal will hold off the pests. The oil of the Australian tea tree, which is notably free of insect pests, is another one mentioned, as is dried wormwood. Adding chopped garlic to your animal's food can also repel fleas. One large clove a day is recommended for a large dog. According to Carl.E. Schreck, formerly of the United States Department of Agriculture's Insect Repellent and Attractant Project in Gainesville, Florida, many Southerners plant wax myrtle near their homes' foundations to repel fleas. Dr. Belfield recommends rubbing ground cloves or eucalyptus oil into your pet's fur as an alternative to an insecticidal collar.

Recently, University of Georgia scientists found that grated citrus peel kills a number of household pests, including fleas. Here is one instance where science has lagged behind folk wisdom. Years ago, Midwesterners repelled crawling insects by regularly scraping a grapefruit rind across the saddles of their front doors. Using the same principle, you can make your own nontoxic antiflea lotion. Here's the recipe:

Cut four lemons in eighths. Cover them with water, and bring it to a boil. Reduce the heat and simmer for forty-five minutes. Allow it to cool and strain out the lemon pieces, reserving the liquid. Store it in a glass container (it may seep through plastic).

Wet your animal thoroughly with the infusion, brushing its coat while wet so that the juice and oil from the lemons penetrate down to the skin. Dry the fur thoroughly with towels, and brush again.

Some pet owners have kept their dogs free of fleas for years by using this homemade lotion regularly.

Traps and Other Mechanical Controls

People have been applying their ingenuity to the problem of flea control for a long time. Someone in the eighteenth century even devised a human flea trap to be worn around the neck. (It probably didn't work.) A more recent trap, needing an accommodating woman, is based on the female sex's reputed attractiveness to fleas. A woman wearing slacks smeared with a sticky substance—petroleum jelly, for instance—walks through a flea-infested room. The insects, aroused by her warmth and body vapors, jump onto her clothes, and become stuck. There are a number of other mechanical controls you may wish to try as well:

❑ Some African villagers reputedly put a lighted candle in a dish of water, which they place in the middle of their huts to attract fleas. You can modernize the idea with a flexible-necked lamp. According to an article by Jo Frohbieter-Muellers that was published in *Mother Earth News,* you should set the lamp next to a shallow dish filled with water and detergent, placing both lamp and dish near an infestation. The detergent will eliminate the surface tension that could turn the water into a flea trampoline. Leave the light on all night for at least a month. The fleas will jump toward the light's warmth, and fall into the water. Of course you have to be careful to keep the water dish a good distance from the electricity of the lamp.

❑ There are a few effective high-tech electronic flea traps. One uses a pulsating flashing light to attract fleas and a sheet of sticky paper to trap them. The University of California, Riverside, reports catching hundreds of fleas in an infested room over a period of several days.

❑ More reliable than either the woman-in-slacks trap or the lamp-and-water trap, and less expensive than an electronic trap, is the flea comb, available in pet stores. When combing your pet, have a container of *soapy* water nearby for drowning your catch. Be sure to get the comb's teeth all the way down to the animal's skin. Petroleum jelly can keep the trapped insects from jumping off the comb. Either rub the petroleum jelly on the comb's teeth now and again while combing, or zap the fleas with a cotton swab dipped in the jelly. Since the fleas may be carrying harmful organisms, you should not crush the insects with your fingernails.

❏ Regular, frequent bathing in summer can control fleas on both dogs and—that rare jewel—a cat who doesn't mind a dunking. Since fleas drown easily, any gentle shampoo will do the job. If your pet objects strenuously to getting wet, try a sponge bath with denatured alcohol, which may be mixed with vinegar; or sponge your animal with a strong brew of wormwood tea. Wormwood leaves are available from commercial herbalists.

The insecticides that promise so much can prove to be less than adequate in controlling fleas, even if you don't mind the exposure to the toxic products. People have had their houses and yards treated and their animals dipped in organophosphates—only to have the pests soon come back.

This illustrates why diligence is so important in flea control. When you find the combination of environmental controls, repellents, and mechanical controls such as baths that works best for you and your pets, these should do at least as well as the chemicals. However, they are not one-shot solutions; you must apply or use them regularly and frequently in warm weather.

For the long haul, look into improving your pet's nutrition. Some veterinarians who promote high nutrition for dogs and cats claim it can increase an animal's resistance to fleas.

If you really cannot tolerate any fleas at all, you should probably find another home for your pet. Or you might consider adopting a parakeet instead. Parakeets don't have fleas.

TICKS

For city dwellers and suburbanites who have dogs, ticks are not much of a problem. Cats, possibly because they lick themselves so frequently and thoroughly, also rarely get ticks. You can have a dog for many years and never encounter this pest. But as with all other forms of wildlife, as we move closer to wilderness areas, or acquire vacation homes in mountains and forests, we become more likely to meet up with this relative of spiders, mites, and scorpions.

Fortunately, ticks are not nearly as invasive as fleas. Except for one species, the brown dog tick, they are normally found only on grass or bushes, and do not become established in homes or kennels.

The Hazards Of Ticks

Ticks run a close second to mosquitoes in the number of diseases they transmit to humans. Some species of ticks are dangerous because they can transmit microbes that may cause paralysis, Rocky Mountain spotted fever (which, despite its name, is found mostly in the southeastern United States), encephalitis, tularemia, and relapsing fever. The bites of all ticks can be an uncomfortable nuisance and, if neglected, may lead to serious complications.

A dog infested with ticks can be greatly weakened by loss of blood. The adult parasite is most often found on a dog's ears and neck and between its toes. Larvae and nymphs (the young adults) settle on the long hairs along the animal's back.

The American dog tick can transmit Rocky Mountain spotted fever; the brown dog tick does not. If, after looking at Figure 8.2, you are still not sure which species is feeding on your dog, take a few specimens in a jar to your local farm bureau or health agency for positive identification. It is safest to handle the specimens with tweezers to avoid contact with any disease organisms they may be carrying.

One tick-borne disease, first identified in the United States in1985, has been named Lyme disease after the Connecticut town where attentive mothers first noted a cluster of cases of childhood arthritis. The deer tick was fingered as the culprit that transmits this illness. Actually, early in the twentieth century, European doctors observed a slowly spreading rash on patients and advanced the idea that it was caused by a tick. In the 1940s, physicians observed that the rash often developed into a multisystem illness. In 1969, a Wisconsin physician successfully treated with penicillin a case of what would later be called Lyme disease. At least twenty-seven different species of birds can carry deer ticks in their immature tick stages; migratory birds have brought this tick and its infectious organisms to the Midwest and the West Coast. It is often misdiagnosed as influenza, rheumatoid arthritis, syphilis, Lou Gehrig's disease, even Alzheimer's disease.

The U.S. Centers for Disease Control (CDC) first began keeping records of Lyme disease cases in 1982, and it became a nationally notifiable disease in 1991. In 1988, about 5,000 cases of this disease were reported in the United States. By 1999, that total had climbed to more than 16,000.

Figure 8.2
Brown Dog Tick
(left); and American
Dog Tick (right)
*Source: Common
Ticks Affecting Dogs*
(Berkeley, CA: Division of
Agricultural Sciences,
University of California,
Leaflet 2525, 1978,
revised).

BROWN
DOG TICK

AMERICAN
DOG TICK

If you find a tick feeding on you or your child, remove it as soon as possible (see page 148) and take both your child and the tick to your doctor. The physician can have the parasite identified and, if necessary, can begin treatment for any potential tick-borne disease.

SHUT THEM OUT

As with many pests, prevention is your best weapon. Following are some suggestions:

❑ On returning from a walk with your dog in a rural area, check the animal for ticks. When the parasites first attach themselves, they are fairly easy to remove and have had little chance to pass disease organisms to the dog.

❑ Rodent-proof your home, especially if it is a mountain cabin, to keep out tick-bearing rats, mice, squirrels, and the like. (See Chapter 5.)

❑ Keep your shrubs and lawns closely trimmed so that the parasite and its rodent hosts find no hiding places near your home.

❑ If you go hiking in the woods, wear sturdy pants, boots, and a long-sleeved shirt with buttoned cuffs. Tuck your pants cuffs into your boots, and avoid sitting on logs in brushy areas. Some experts recommend wearing hard-finished (that is, impenetrable) light-colored clothing so that you can easily spot any ticks. Be aware that ticks can crawl under your clothing to your skin, so frequent inspection of your skin can help you get rid of them before they do you harm. If you are out hiking alone, carry a hand mirror with you so that you can inspect your armpits, groin, back, and head.

❑ Before choosing a picnic spot in the woods or on the grass, draw a piece of light-colored cloth over the area. Any ticks lurking there will be drawn to the cloth, and you will know to choose another spot for lunch. Or if you are in an unfamiliar place, check with park or recreation authorities. They should know which areas are infested and which are not.

❑ If you do a lot of hiking, include a "tick tweezer"—available in sporting goods stores—in your gear, so that you will be able to remove the pest before it can do you any harm.

❑ When hiking in a heavily overgrown area, stay near the center of the trail. When you return, examine your entire body for ticks, and ask someone else to check your head and back. Ticks have a habit of wandering over a host's body before settling in for a blood meal. In some regions, both tick adults and nymphs are disease-carriers. Nymphs live within and beneath leaf litter, so any activity that puts you in direct contact with leaves or evergreen needles—such as gardening, picnicking, sitting or lying down on the ground, or sitting on a log—may increase your risk of being bitten.

❑ You can spray commercial repellents on your skin or clothing to deter ticks; however, since water or perspiration can wash off repellents, you may need to apply them often after swimming or perspiring heavily. To make life even more complicated, if you are fully clothed, a tick may attach itself to your scalp or behind your ears.

WIPE THEM OUT

Since, as noted above, ticks rarely cause the type of infestation that fleas do, the main focus of wiping ticks out is checking for them and, if you find one, removing it. It is important to remove ticks safely. To do this, attack them as you would a large splinter, using a pair of tweezers and an antiseptic. Grasp the animal by its tiny head, as close to the skin as possible. It is most important that the head not be broken off and left in the wound. If this happens, ulcers, infection, even blood poisoning can result. Also, try not to crush the tick while pulling at it. *The Complete Home Medical Guide* recommends using a steady rocking movement while pulling with the tweezers. Despite the folk wisdom, *do not* apply a lighted cigarette or other heat source to the animal as this can cause it to vomit or salivate, releasing any disease organisms it may be carrying. Although a tick has to be embedded for two or three days to transmit Lyme disease, a tick carrying Rocky Mountain spotted fever can start its dirty work on Day One.

Coating the pest with petroleum jelly may make it withdraw on its own within a half-hour to keep from suffocating. Entomologist Karl von Frisch recommends applying fresh adhesive tape if the tick is not yet very swollen. Perhaps the chemicals in the adhesive irritate the pest, or perhaps the tape simply suffocates it. At any rate, when you remove the tape the next day, the tick will come away with it.

After removing the tick, you should always apply an antiseptic to the site of the bite, just as you would to other wounds. If your hands have touched the tick during removal, wash them thoroughly with soap and water, since the tick secretions may be infective. Whenever a tick is taken from a person, it should be saved in alcohol for later identification in case the person gets sick within *a month* of being bitten.

Now for some good news: Two studies in the July 12, 2001 edition of *The New England Journal of Medicine* suggest that anxiety over Lyme disease may be much worse than the disease itself. It is difficult to catch this disease, and even if you do, it is easily treated. A third study reported in the same journal states that long-term antibiotic treatment does nothing to cure Lyme disease and may cause serious harm to the patient, including destruction of bone marrow.

"If you go long enough without a bath, even the fleas [and ticks] will leave you alone," wrote correspondent Ernie Pyle during World War II. Though certainly a nonchemical control, going bathless hardly seems a socially desirable technique for repelling fleas and ticks. Such a tactic is more apt to repel your friends. Whether you have a dog or cat or just enjoy a woodland ramble, the strategies described in this chapter will keep you and your pets free of these dangerous parasites—without destroying your social life.

9. Lice and Bedbugs— The Unmentionables

Though usually linked to personal filth and dirty surroundings, lice are really democratic, and will readily move from one individual to another no matter how fastidious the new host, nor how well placed on the social ladder.

Bedbugs are equally undiscriminating. Not long ago, a family whose home boasted a grand piano and Oriental rugs bought an almost-new sofa bed from acquaintances. The first night the family's son slept on it he was badly bitten, and the next morning there were itching welts all over his body. The bed was quickly thrown out.

It's easy for the unwary to pick up either of these nuisances, and it's very difficult to get rid of them. First, let's deal with the more dangerous of the two—lice.

LICE

Lice are wingless round, flat insects from one to four millimeters long. They are grayish-white (crab lice may have a pinkish tinge), and have three sets of short, stout legs. One or more of the legs have a claw that is well suited to grasping a human hair. Pubic (crab) lice are shorter and broader than head lice and have two pairs of heavy claws.

Head and pubic lice spend their entire lives on the host's body. Body lice, by contrast, hide in a person's clothing when not feeding. Because humans are warm, lice breed all year long. Crabs and head lice cement their eggs to body or head hair; body lice attach them to the fibers of underclothing, especially along the seams and at the neck, shoulder, armpit, waist, and trouser crotch. With the necessary warmth, eggs hatch in about a week. Temperatures above 100°F and below 75°F either greatly reduce hatching or prevent it completely. Depending on the species, the insect lives from ten days to about a month.

This parasite cannot survive if the host's body temperature drops because of death or rises from fever. However, it can readily change hosts if there is close personal contact. Off the person, all life stages of lice die within thirty days, regardless of temperature.

Lice move very rapidly, especially when exposed to light. Contrary to what you may hear or observe, they neither jump nor fly. However, static electricity in a comb can expel the creature suddenly, giving the impression of a leap or flight.

Lice can move directly from one host to another. *You cannot get lice from an animal or insect.* These parasites are passed only from one person to another through direct contact or shared clothing, bed linens, towels, and upholstery.

You don't *have* to be personally filthy or live in squalid surroundings to catch lice, but it helps. Once you have lice, filth will guarantee their increase, and crowded conditions will spread them.

Species of Lice

There are three basic types of lice: body lice, head lice, and pubic lice (also known as crabs). (See Figure 9.1.) While there are similarities, there are also differences.

Body Lice

Body lice are picked up through direct contact with an infested person's body hairs or clothing. They also live in bedding used frequently by an infested person, and can move to the bed's next occupant.

Body lice peak in winter in cold climates, where people wear several layers of clothing. Typhus, which is carried primarily by body lice, is endemic in such regions in the cold months.

The classic symptom of long-term infestation with lice of any type is scarred, hardened, or pigmented skin. But long before things get to that stage, you can easily spot telltale signs. For body lice, these include dirty brown patches of their droppings around the armpits, scabs, excessive scratching, and inflamed, irritated skin. Body lice seen on outer clothing signify that there are likely many more on inner garments.

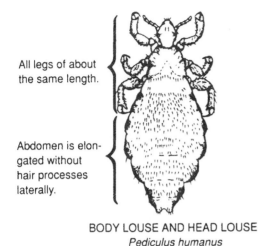

All legs of about the same length.

Abdomen is elongated without hair processes laterally.

BODY LOUSE AND HEAD LOUSE
Pediculus humanus

Figure 9.1
Body Louse, and Head Louse (top); Crab Louse (bottom)

Source: Lice of Public Health Importance and Their Control (Atlanta: Centers for Disease Control, Homestudy Course 3013-G, Manual 7-B, 1982).

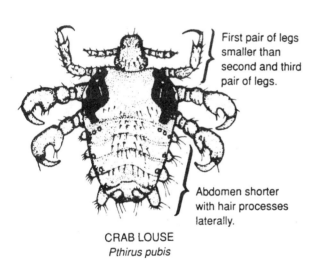

First pair of legs smaller than second and third pair of legs.

Abdomen shorter with hair processes laterally.

CRAB LOUSE
Pthirus pubis

Head Lice

A few years ago, soon after fall classes started at an elite women's college in the Northeast, a student asked a classmate, with whom she had just spent the summer working at a children's camp, to come to her room. With the door closed, she asked, "Do you know lice when you see them? I think I've got them."

Lifting locks of clean short brown hair, the friend spotted one or two tiny grayish-white creatures scurrying away from the light.

"I think you do too," she said, "and I think these are nits," she added, pulling off little yellowish ovals from a few individual hairs.

This unlucky young woman spent the next two weeks in the college infirmary, her head frequently wrapped in a towel.

In another instance, a smartly dressed matron from one of the Los Angeles area's exclusive suburbs brought her daughter to an upscale hair salon for a shampoo, haircut, and styling. The owner of the shop took one look at the girl's thick blond locks and said, "I can't touch her. The Health Department would shut me down if I did. Her head is crawling with lice." The enraged mother stormed out.

You can be finicky about yourself, your friends, and your environment, and still contract head lice (or pubic lice), although such bad luck is more often the result of poor personal hygiene or loving contact with someone who has them. Head lice can also be transmitted by sharing such items as combs and brushes, wigs, hats, athletic headgear, earphones, and towels with an infested person. Probably the most common species in the United States, head lice most often infest school-age children.

Signs of head lice include the presence of brown fecal matter on the person's shoulders and back, which is easily seen on light-colored clothing. Louse eggs, or nits, may be seen along the hairline, behind the ears. There may be more head-scratching than usual, difficulty sleeping, and irritability.

An individual with tiny yellowish globules on the hair, especially behind the ears at the hairline, but with no visible lice, may or may not have an infestation. The globules could be hair spray or dandruff. To be certain, try to slide them along the hair shaft. Dandruff and hair spray will slide right off, but nits, which are cemented firmly in place, will not.

According to the National Pediculosis Association, 6 to 12 million school children catch head lice every year. Lice are more widespread than any communicable childhood disease, with the heaviest transmittal occurring during the first two months of school.

In addition, adults often assume that school-age children are able to care for their own hair, and therefore do not supervise shampooing. The youngsters themselves don't recognize the real nature of their problem. They just scratch.

Have head lice heard about affirmative action? African Americans are generally free of this nuisance. The reason seems to be that our native pests can only grasp round hair shafts, while those of African Americans are generally square. Head lice in Africa have adapted to the square hair shaft and folks on that continent are not spared this botheration. People from Southeast Asia are also relatively free of head lice. Many use a type of hair dressing based on coconut oil, which is deadly to lice.

Pubic Lice

Pubic lice (also known as crabs) are most often passed during sexual contact. Telltale signs include dark spots on the person's underwear or around the armpits, as well as itching, which may be severe at times. There may be occasional painless blue spots on the skin, something not seen with head lice.

Like venereal disease, the occurrence of pubic lice increased greatly with the sexual revolution. But even people with a circumspect sex life, or no sex life at all, can attract these unpleasant, embarrassing little pests, since they can infest public toilet seats and beds, as well as towels and clothes in gymnasium locker rooms or wherever many different people's clothes are crowded together. They can also be acquired from people's stray body hairs.

Although they usually lie in the armpit, pubic, and perianal areas, crabs can also infest beards, moustaches, eyebrows, eyelashes, and chest hair—wherever hairs are spaced fairly wide apart. On infants, who can pick them up from their caregivers, these parasites can live on the eyelashes and eyebrows and along the hairline. If you find them on a young child, suspect sexual abuse.

The Hazards of Lice

Lice, like mosquitoes and fleas, are carriers of diseases that have killed millions. They are the inevitable accompaniment of war and other catastrophes that prevent people from changing their clothes or bathing frequently.

Louse-born typhus (more deadly than murine, or endemic, typhus, which is carried by fleas) has at times been more important than military skill and strength in deciding a war's outcome. In part because typhus decimated the French Army in 1528, the Spanish defeated them, and Spain dominated Europe for the rest of the century.

During World War I, typhus and trench fever, also louse-borne, killed German and Allied soldiers evenhandedly. In World War II, typhus raged in North Africa, the Balkans, Russia, Italy, and the Nazi concentration camps, killing soldiers and civilians alike.

But World War II also brought a breakthrough in the control of typhus and the parasite that carries it. Dichloro-diphenyl-trichloroethane (DDT), a pesticide then newly invented in Switzerland, was dusted on people during an incipient typhus epidemic in Naples, and the disease was halted. For the first time in history, a typhus outbreak was stopped in winter. Today, typhus can be cured by antibiotics. But the lice live on.

Although highly contagious, head lice are not known to carry any dangerous disease. The constant head-scratching they cause, however, can tear the scalp's skin, leaving a person open to the same secondary infections as any other break in the skin. But this pest is not nearly so dangerous as the pesticides often overused by anxious parents who do not know that there are safer routes to a cleanup. And head-scratching, with its possibility of broken skin, leaves the child vulnerable to the toxins. Besides that, an overanxious parent applying a strong pesticide to a child's scalp also is being exposed to the poison.

SHUT THEM OUT

As with most parasites, both human and animal, it's simpler to prevent an infestation of lice than it is to clean one up. Different preventive measures are useful for different types of lice.

Body Lice

❑ To protect yourself against body lice, bathe and change your outer clothes frequently, and your underwear daily.

❑ Don't go without underwear. Since it is changed and laundered often, just wearing underwear regularly is a good preventive.

❑ Avoid wearing borrowed clothing; you could be borrowing lice.

❑ When acquiring used garments, either from an acquaintance or a thrift shop, launder them or have them dry-cleaned before wearing them. Or store the items for four weeks in a plastic bag to kill any lice and their young.

❑ Avoid sleeping in beds used by people who may be infested. If you must use such a bed, make sure the linens and blankets are fresh.

❑ Inspect your body and clothing frequently if you think there is a chance of infestation.

❑ Avoid contact with infested people, especially at night, when the parasites are more active.

Head Lice

Whether you are rich or not so rich, preventing head lice is a matter of a few simple procedures carried out consistently:

❑ As early as 1913, it was recommended that girls braid their hair before going to school, and unbraid it and comb it out when they come home. Braiding kept the hair from making contact with that of other children, preventing it from flying about and possibly picking up a few lice from a playmate. This is still a good precaution. Today's long, loose locks, bouncing around with the wearer's every move, can readily pick up an infestation from another nearby head with long, loose, lousy locks. Being stylish is one thing; using common sense is another. Another common practice a century ago was to cut boys' hair very close to their heads. The boys may have looked like skinned rabbits, but they didn't catch head lice.

❏ Another sound preventive measure is regular shampooing with a coconut-oil–based shampoo. Coconut oil is the source of dodecyl alcohol, which is lethal to adult lice and some eggs. Green soap, once the standard shampoo at summer camps and other institutions, is based on coconut oil.

❏ Because the transmission of head lice most often occurs at school, schools must play an active part in preventing the spread of these parasites. Schools usually have a strong interest in dealing with a head-lice problem. Many school districts have policies mandating that a child infested with head lice be kept out of school until he or she is louse-free. That translates into absences, which means a loss of state revenue until the youngster can return to school. Obviously there can be a huge economic impact to a school district if there is a widespread epidemic of the parasite. Complicating the problem is that schools in major cities often have students from several different ethnicities in the same classrooms, and some cultures do not consider an infestation of head lice to be much of a problem. With a "No-Nits" policy (which requires that any child found to have head lice be sent home and be examined for lice before being readmitted) in force, it may take some persuading to get parents of different backgrounds to go along.

❏ If you have school-age children, insist that school authorities provide individual storage for students' clothing. Head lice spread quickly if children store their clothing haphazardly. Youngsters with assigned hooks and/or individual lockers have fewer cases of pediculosis than those who share clothing storage.

❏ Make sure that any day-care center your child attends does *not* have a dress-up corner with shared hats, scarves, and other clothing.

❏ Don't assume that your child's school is so sanitary that it can't possibly harbor this pest. When the city of Indianapolis screened its schools, health authorities found head lice in every school, and some infestation rates were as high as 18 percent. For help in establishing an effective louse-screening program, contact the National Pediculosis Association (50 Kearny Road, Needham, MA 02494; telephone 781-449-6487; website www.npa@headlice.org).

❏ If your child's school is suffering an outbreak of head lice, inspect and comb your youngster's head each day right after school. A comb warm enough to be pleasantly hot to the hand will rout the lice from their hideouts among the hairs, making them easy to remove. Have a hand-held hair dryer nearby to warm the comb from time to time as you work. Be sure to drop *all stages* of the organism that you catch into boiling water to kill them and their disease germs. Don't crush them with your fingernails, as this will release their harmful germs onto you.

❏ During a school epidemic of pediculosis, check every other member of the family, since these parasites move very quickly from person to person in close quarters. If you are inclined to shrug the whole problem off, remember that a neglected case of head lice can produce a foul mass on the scalp, one in which a fungal infection can readily develop.

Pubic Lice

To prevent pubic lice, avoid contact with infested people. This may be harder to do than it might seem. In 1999, the management of the squeaky-clean Magic Kingdom settled a dispute with employees whose jobs involved wearing the costumes of Mickey and Pluto and the rest of the gang. The actors were picking up crabs and scabies from the garments worn under the animal costumes. Although the clothing was washed in Disney facilities, evidently the pests were still active. It took two months of negotiations to win the employees' right to take the undergarments home and wash them themselves.

Following are commonsense steps to keep from preventing pubic lice:

❏ If no paper cover is provided in a restroom, turn the toilet seat up for a few seconds before using it so that any lurking parasite slides off.

❏ Don't share gym towels or lockers.

❏ Make sure that any strange bed you sleep in has fresh linens.

❏ Consider the possibility that a new sexual partner may have crabs.

WIPE THEM OUT

By the time lice are noticed, they cannot be shut out as they are already feeding on their unwilling host. But they can be wiped out . . . and without resorting to chemical pesticides.

Since lice do not easily drown, soap and water at temperatures we can stand don't get rid of them. That's why, in many parts of the world, people who continually wash their bodies may still have lice. Following are measures recommended to get rid of lice.

Body Lice

Exterminating body lice without resorting to the use of toxic chemicals is simple, and involves two main elements: proper laundering practices and personal hygiene.

A person infested with body lice who frequently changes into properly laundered clothing will eventually get rid of the parasites without any other treatment. Proper laundering means agitating all linens and washable garments in *hot* water, at least 140°F, for twenty minutes. If circumstances are such that you cannot wash the clothes, store them in a sealed plastic bag for thirty days. This will kill any lice on the garments. The University of California Cooperative Extension Service recommends washing clothing and bedding with a 5-percent creosol solution, or soaking them for thirty minutes in a 2-percent solution of the substance, though this is optional.

Woolen clothing belonging to an infested individual should be dry-cleaned. Cleaning solvent is lethal to lice, as is the heat of pressing. Put the suspect garments in a tightly closed bag before taking them to the shop. Not every cleaner will accept infested clothing, especially those who do not operate their own plants, so check around first. The U.S. Centers for Disease Control and Prevention says that just pressing woolen garments along the seams gives good control, since this is where body lice lay their eggs. Store clothing in plastic bags for thirty days.

In addition to proper laundering practices, it is vital to maintain good personal hygiene by bathing frequently and changing underwear daily.

Head Lice

Before you panic when your child comes home with a note from the school nurse saying that your child is infested with head lice and will not be allowed to return to school until free of nits and lice, calm down. (Most schools have a "No-Nits" policy that requires that any child who has been determined to have head lice be re-examined before being readmitted to school.) Take the time to inform yourself of the realities of what is really a minor health problem, and learn how to deal with it as rationally as you would with a mild cold. First, realize that an infestation of head lice is no reflection on you or your personal hygiene. Head lice are no respecters of socioeconomic class or household hygiene. It is no disgrace to catch head lice, but it is a disgrace to keep them.

The quickest, cheapest safe control for head lice is a drastic haircut, ¼ inch or shorter, or shaving the head. Either of these could bring painful ridicule on a child, however. There is another, less traumatic alternative, one proven effective for generations. This is a combination of shampooing, soaking, and fine combing. Pediculosis expert Benjamin Keh recommends a shampoo, followed by a soak in the lather and then a meticulous combing with a comb specially designed for this job.

First, get a LiceMeister comb. This sturdy steel-toothed comb was first developed in Germany in the early twentieth century, and is extremely effective in scraping off nits. It is the only comb recommended by the National Pediculosis Association (NPA). You can have your local pharmacist order it, or buy it directly from the NPA. Major drugstores also carry it. It costs about ten dollars. Here is Dr. Keh's traditional method for cleaning an infested head. Although admittedly tedious, it will work. More important, your child will be in no danger from potentially toxic pesticides. And you and your youngster might even enjoy the quiet time together.

1. Wet the hair thoroughly with warm water.

2. Apply a soap shampoo (preferably coconut-oil–based), and work it into a thick lather. Rubbing the scalp and hair thoroughly, start from the back of the neck, the ears, and the forehead, and work toward the center of the head. Be sure that the entire head and all the hair are covered with shampoo. (You can use vegetable oil instead of shampoo, but then you will need to wash the hair twice after the combing—once to clear off the oil, and a second time to clear out the dead insects.)

3. Brush this lather off, and rinse again with warm water.

4. Repeat lathering and rubbing, this time leaving thick suds on the hair. Tie a towel around the lathered head, and leave it on for a half-hour.

5. Remove the towel. The hair will be soapy. *Leave it this way.*

6. Comb the hair with a regular comb to remove tangles.

7. Separate a one-inch strand of hair. Hold the LiceMeister comb with the beveled side toward the head. Lifting the strand of hair away from the head, comb the strand slowly and repeatedly from the scalp to end of hair until every nit is removed. (If the comb misses any nits, remove them with your fingernails and drop them into boiling water.)

8. Pin the clean strand out of the way. Start the next strand and comb in the same way. Continue combing strand by strand until the entire head is free of nits. *Important:* If the hair dries during combing, wet it with water. Wipe nits off the comb frequently with cleansing tissues; drop the tissues into a plastic bag. It usually takes two hours of constant combing to remove all nits, longer if the hair is thick and long.

9. After removing all the nits, lather the hair again, rinse, and dry.

10. When the hair is thoroughly dry, inspect the entire head for any missed nits and remove them.

11. Following the shampoo and combing, immediately place the comb in hot, soapy water to which a couple of tablespoons of ammonia have been added. Allow the comb to soak for at least fifteen minutes. Then scrub the teeth of the comb with a stiff nail brush. There is a specially designed cleaning tool that comes with the LiceMeister comb, but you can also remove any dirt lodged between teeth by pulling dental floss through the openings. *Do not* leave any dirt on the comb's teeth. Boil the towels and brush after this procedure.

12. Repeat hand-picking and/or combing each day for twelve days.

Inspect and fine-comb the hair daily until the epidemic is over. Since lice can be transmitted through upholstery, vacuum sofas and chairs every day until the infestation is over. Just running infested items through a cycle in a hot-air dryer will help. Take special care to launder all clothing frequently and to dry-clean non-washables. As with body lice, storing clothes in plastic bags for four weeks will also kill the insects.

Because there is no substance that can dissolve the cement that glues louse eggs to the hair without damaging either hair or scalp, the method described above is the safest, most effective way to deal with pediculosis. A slightly less tedious method is to use hair conditioner instead of a shampoo. However, conditioner only slows down the lice for about twenty minutes, which may not be enough time for you to do a thorough job. You can also apply salad oil to the hair, which has the advantage of not drying out during the combing. But then it will take two shampoos to get rid of the oil. Any method will be time-consuming, *but your child will be safe.*

Folk wisdom holds that vinegar will dissolve the nit glue. It doesn't work. Unhatched eggs are usually ¼-inch or less from the scalp. Any that are farther down the shaft have probably already hatched. Health authorities disagree about the exact distance. In the interest of safety, it's best to remove all the nits.

Pubic Lice

The simplest treatment for crab lice, says the Centers for Disease Control and Prevention, is the same as that for head lice:

❑ Shave *all* body hair very closely, eliminating all stages of the pest.

❑ To guard against reinfestation, boil all underwear, bedding, and towels for the entire household.

❑ In young children and infants, apply 0.25 percent physostigmine eye ointment with a cotton swab to eyelashes and eyebrows, and along the hairlines. This ointment can be bought only with a doctor's prescription.

❑ Although it has been used for generations, mercuric ointment is neither effective nor safe. Its value as an insecticide is low, and it carries serious risk of poisoning.

BEDBUGS

"We get calls from people in Beverly Hills," says Corinne Ray, formerly director of the Los Angeles Poison Information Center. "They ask, 'What should we do about all the spiders in our bedroom? We're being bitten all over.'"

Spiders bite only when they feel threatened, she tells her callers, and then only once. "You probably have bedbugs," she adds. The callers abruptly hang up.

Few people like to admit they have bedbugs, because the pest is so often associated with dirty homes and poor hygiene. But a generation or so ago, they were quite common, especially tormenting travelers even in "respectable" hotels.

In the 1930s, there was a great surge of bedbugs in northern Europe, brought on, it is believed, by the rapid adoption of central heating. The problem was so bad in Sweden that a number of cities seriously considered building special hotels to house people whose homes were being fumigated. Tents, used in summer, were out of the question in Sweden's cold winters.

Around the same time, about 4 million Londoners were living in bedbug-infested homes. Fumigation of furniture was almost compulsory, and the homes themselves were thoroughly treated between occupants.

Using Pesticides: A Drastic Measure

On September 5, 2001, then-Governor of California, Gray Davis, signed a bill outlawing the use of the pesticide lindane for treating lice and scabies. This was the culmination of a decades-old effort by the National Pediculosis Association, Consumers Union, and Public Citizen to keep this highly toxic substance off drugstore shelves. "It was with L.A. Sanitation District's documentation of excessive amounts of lindane in Los Angeles County's water supply that the serious nature of the chemical was brought to the forefront and responsibly addressed," said the National Pediculosis Association. "It's not about lice," is their watchword. "It's about kids."

Sold under various commercial names, lindane is the insecticide most widely used to clean up head lice. Until 1983, the federal government severely restricted its use against organisms other than head lice, and with good reason. Lindane can be absorbed through the skin, and has caused convulsions, seizures, and cancers in laboratory animals. One child is reported to have died from lindane poisoning after treatment for head lice, and another had a seizure. The insecticide has also been implicated in several cases of a grave form of anemia.

Even if lindane is still sold where you live, here is a story that may give you pause before you rush off to the drug store for this insecticide. The mother of a two-year old diagnosed with scabies (a rash caused by a mite) applied a prescription containing lindane to the boy's legs. The pediatrician had assured her it was safe, saying, "I've used it for years." Some of the rash had open sores. Following the doctor's instructions, the mother waited ten days before applying the lotion again. After a third application of lindane to the child's skin, he had a seizure, and on the way to the emergency room, he stopped breathing.

He was revived, but the seizures continued. The boy, now eight years old, has cerebral palsy and a damaged left foot, leg, and arm. The Public Citizen's Health Resource Group has had almost fifty reports of convulsions from using products containing lindane. Nearly half of these cases involved children under the age of ten.

Admittedly, lindane has been used many times without any reports of ill effects, but that will make little difference to the parent whose child is the exception. In addition, lice are becoming resistant to the chemical, so exposing your child to this powerful substance may not clear up the infestation. Like many other chemicals we have been using so freely over the past few decades, lindane's long-term effects are not yet completely understood.

Incidentally, veterinarians used to use a lindane solution of 0.03 percent to kill fleas and ticks on small animals. Yet a widely sold lindane-based shampoo recommended to get rid of head lice is mixed at a strength of 1.0 percent. Are children presumed to be less vulnerable to this powerful toxin than their pets?

Adding to the difficulty with head lice is the growing resistance of the pests to all the insecticides we have been throwing at them, so panicky parents who clear drugstore shelves of one questionable product after another are putting their children in harm's way without solving the problem. Doctors at the University of Miami medical school recently reported that lice they collected could "crawl around for hours without slowing down at all" on cotton soaked with permethrin, a safer alternative to lindane. Permethrin is a relatively safe pesticide that is derived from a kind of chrysanthemum. It is still legal in some states. However, chrysanthemums are related to ragweed, so if you have hay fever, you may want to avoid using permethrin.

When you buy a louse comb and shampoo, the salesperson may suggest that you use an environmental spray at the same time. In addition to the dangers that go with releasing any aerosol pesticide into the home environment, should you take this salesperson's advice, you will be using a product whose manufacturer has neither applied for nor received approval from the U.S. Food and Drug Administration. The Centers for Disease Control and Prevention and the National Pediculosis Association both strongly advise against the use of any such spray.

Finally, if your treatments at home do not control the infestation, take your child to a doctor. You should also seek medical advice for anyone who gets a severe reaction to louse bites.

Bedbugs are still a problem, usually under primitive or unsanitary conditions. And when people move from infested housing to clean accommodations, they easily carry the pests with them. Even well-kept university housing, according to the University of California's Professor Walter Ebeling, is not exempt. (Have college students ever been known for their meticulous housekeeping?)

Also called mahogany flats, redcoats, chinch bugs, and wall lice, bedbugs are a hardy breed. Four to five millimeters long, oval, and flat when hungry (see Figure 9.2), the creatures become plump and elongated after a feeding. Under normal conditions, they can live as long as a year, but starvation and cold weather actually prolong their lives. Without food and at tempreatures of 60°F or below, adult bedbugs enter semihibernation and can live through the winter in an unheated building. Eggs and newly hatched young, however, die within a month at subfreezing temperatures, and so do not survive winter in an unheated structure.

Bedbugs invade otherwise clean homes from various sources. Secondhand furniture is the most frequent means of entry, but the pests can come from bat roosts, swallows' nests, and chicken and pigeon coops. Other sources include ill-kept theaters, furniture retrieved from storage, poorly maintained upholstery on public transportation, infested clothing laid on a bed, and moving vans that have carried infested furniture. If an infested apartment is vacated and remains empty, the pests will spread, in time, to nearby occupied quarters.

The Hazards of Bedbugs

Although they have not proven this, scientists say that bedbugs may transmit

MALE BEDBUG   FEMALE BEDBUG

Figure 9.2 Male and Female Bedbugs

Source: Hugo Hartnack, *202 Common Household Pests of North America* (Chicago: Hartnack, 1939).

plague, relapsing fever, tularemia (also known as "rabbit fever"), and Q fever. This last disorder, related to Rocky Mountain spotted fever, was first identified in Queensland, Australia; hence the odd name. Even without the possibility of transmitting disease, however, the bedbug is a most unpleasant visitor. Its bite leaves hive-like welts that itch intensely. Like fleas, bedbugs pierce the skin to suck their host's blood. Also like fleas, the bedbugs' saliva causes more discomfort than the bite. In extremely sensitive people, the swelling caused by the saliva can be severe.

How Bedbugs Make Themselves at Home

Once in a home, bedbugs concentrate on surfaces that are rough, dry, and partially or completely dark. They prefer wood or paper, where their excrement and eggs soon become visible. They like to lay their eggs, firmly cemented in place by a sticky secretion, behind loose wallpaper; in wall and floor cracks, nail holes, and light-switch boxes; and on door and window frames. They will hide in a wooden bed frame, but you may not be safe even with a metal bed frame because the insects readily travel from their hiding places to find a blood meal, even if it is in another room. Some people who sleep in metal beds have been awakened by bedbugs dropping from the ceiling—a most unnerving experience!

This pest gives off a characteristic smell, rather like very sweet raspberries—only, say some, if the berries have been fed on by stinkbugs.

SHUT THEM OUT

Bedbugs are no different from other household pests in that it's simpler to prevent them than to clear them out. Here are some steps to prevent bedbugs from moving into your home:

❏ Inspect any used furniture before buying it, including beds, upholstered pieces, and rough-textured draperies.

❏ Patch all wall cracks, nail holes, and crevices with caulking compound or latex paint. Make sure all window and door frames are completely sealed and free of cracks.

❏ Paste down any loosened wallpaper securely.

❏ Fill all floor cracks. A fresh coat of varnish on a wooden floor helps.

❏ Caulk all cracks and spaces behind baseboards.

❏ See that the paper backing on all framed pictures is securely pasted down.

❏ Make sure your house's foundation is crack-free.

❏ Screen attic vents so that bats and birds can't get in to roost.

❏ Fit flanges on all pipes entering the house.

❏ Explore an early method of control. In Austria in the 1930s, bedbug-proof construction was considered the surest control. This included coating walls from the floor to three-quarters of the height of the room with slick oil-based paint. Instead of a conventional baseboard, hollow molding, concave to the floor, finished the room. There's no reason why this wouldn't work today, although shiny walls may not be your idea of elegant decor.

WIPE THEM OUT

Getting rid of bedbugs quickly without chemicals is difficult. But before resorting to hazardous substances, try some other means first:

❏ For temporary control, the Center for the Integration of Applied Science recommends smearing petroleum jelly around the legs of your bed (assuming the insects aren't hiding in the bed frame, but just migrate there from the wall).

❏ Send your mattress and pillows out to be fumigated.

❏ Vacuum your room and all its accessories (such as picture frames) thoroughly.

❏ Focus a high-intensity lamp on mattress seams and buttons and on wall cracks for a quick kill.

❏ Expose mattresses and bed frames to pressure-generated steam.

❏ Wash bed linens and synthetic blankets in *hot* water. Have wool blankets dry-cleaned.

❏ Heat your house to 115°F for a half-hour. At all stages, bedbugs die at temperatures above 113°F.

❏ If you sleep in a bed with a metal frame, set the legs in soapy water, smear them with petroleum jelly, or wrap them in double-sided sticky tape. Another ploy is to lay duct tape around the walls and bed legs. The tape can be held in place with masking tape.

❏ If you feel that you must resort to chemical controls, the safest choice is probably pyrethrin dust, which should be laid down in all cracks and crevices likely to be harborages. If the insects are coming from the attic or wall voids, entomologist Walter Ebeling suggests applying a silica-gel product such as Dri-Die 67 to these areas. Although silica gel is no more toxic than talcum powder, you should be sure to wear goggles and a respirator when applying it because of its irritating qualities. (See page 45 for a discussion of silica gel.)

❏ There has been much interest recently in the biological control of insect pests, with some notable successes. Unfortunately, bedbugs are not one of them. Their predators tend to be just as undesirable as they are. Who would want to introduce centipedes, straw itch mites, pharaoh ants, or cockroaches into their bedrooms just to get rid of bedbugs?

Although you may be fastidious about your personal hygiene and meticulous in caring for your home, you can still inadvertently attract lice or bedbugs. As with all household pests, watchfulness is your first weapon against these unwelcome visitors. Getting rid of them without endangering yourself will take some effort and attention, but your safety and comfort demand no less.

PART FOUR

Pests of Property

10. Clothes Moths, Carpet Beetles, Silverfish, Firebrats, and Crickets

Clothes moths, carpet beetles, silverfish, firebrats, and crickets can rob you of as much as a clever burglar. Unlike burglary, however, it's next to impossible to get insurance against insect damage. Your best protection is being aware of what these pests need, and eliminating those conditions.

CLOTHES MOTHS AND CARPET BEETLES

Since clothes moths and carpet beetles wreak much the same kind of havoc, I will consider them together.

In nature, clothes moths and carpet beetles are useful, eating the remains of dead animals that other scavengers leave behind—hair and feathers, claws and nails, horns and hoofs, all of which are composed mostly of keratin, a protein these insects can digest. By using hair and other animal products in our clothing and home furnishings, we have created an ideal environment for fabric pests.

At one time, sheep were dipped in solutions of either lindane or DDT to prevent insect damage to woolen items. However, because of the problems these poisons pose to the environment, this is no longer done. Therefore, any woolens you have need special care to keep these tiny marauders off.

Clothes Moths

Moths are among the most beautiful insects on earth, often sporting gorgeously patterned wings. Though butterflies get all the publicity (there are even butterfly zoos!), there are many more species of moths, most of them harmless night fliers. However, two species are the villains who mutilate our clothes and possessions. In the United States, the *webbing clothes moth,* usually seen in cooler areas, is the most common; the *casemaking clothes moth,* generally found in southerly regions, does the most damage. Equally destructive species are found in other parts of the world. Table 10.1 shows how casemaking and webbing clothes moths differ. Figure 10.1 illustrates an adult clothes moth.

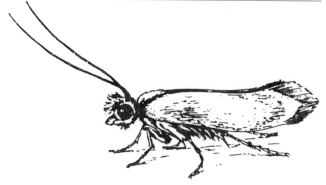

Figure 10.1
The Clothes Moth
Source: Karl von Frisch, *Ten Little Housemates* (Oxford, England: Pergamon Press, 1960). Used by permission.

Female clothes moths lay about forty to fifty eggs on any susceptible material (see page 174) over a two- to three-week period. The eggs hatch three to four weeks later, resulting in larvae that are ready to do their damage. If you are not alert and neglect your woolens, you may meet some flying adults ready to start laying more eggs when you need to use your warm clothes.

Clothes moths are not drawn to light, and hide when disturbed; females heavy with eggs are weak flyers. The eggs of clothes moths can hatch under water, and the larvae survive there for twenty-six hours. One species, when fully grown, can gorge on mothproofed fabrics with no ill effect. The average larva will eat eleven or twelve times its weight in wool. It can starve for more than eight months and still become a fertile adult. This tiny creature is tough!

Carpet Beetles

Carpet beetles, with the exception of the black common carpet beetle, are beautiful and colorful, about the same size and shape as the helpful lady beetle (ladybug). If it didn't move, the furniture carpet beetle—mottled black, white, and golden brown—could be mistaken for a semiprecious stone. The varied carpet beetle is a little less showy, running more to subdued grays and blacks. (See Figure 10.2.)

But don't be deceived by good looks. These insects are the most destructive fabric pests in the world—and the most difficult to control. Their larvae are relatively resistant to pesticides, which is reason enough to use nonchemical controls.

Adult carpet beetles, also called buffalo beetles, are one-sixteenth to one-quarter inch long. Outdoors, they feed on the pollen of flowers, especially white or cream-colored ones. Indoors, they fly to window light and, unlike the shy clothes moths, roam freely about a house. Their sturdy larvae, about one-quarter inch long and shaped like stubby, bristly carrots, move very fast.

The female carpet beetle lays her eggs in dark protected crannies, behind baseboards, in floor cracks, and inside hot-air ducts. The eggs hatch in one to two weeks. From egg to adult may take three years, but in a centrally heated building,

TABLE 10.1 CHARACTERISTICS OF WEBBING AND CASEMAKING MOTHS		

The table below summarizes key characteristics of the webbing moth and the casemaking moth, the principal types of clothes moths you are likely to find in your home. Note that there are many similarities between the two types, but there are also key differences.

Characteristic	Webbing Moth	Casemaking Moth
ADULTS		
Appearance		
Color	Golden buff	Drab buff
Head hairs	Brownish	Light-colored
Length	About $1/4$ inch	About $1/4$ inch
Wings	Satiny	Have vague dark spots
Wingspread	About $1/2$ inch	About $1/2$ inch
Abilities	Similar for both species:	
	• Can fly considerable distances unless carrying eggs	
	• Can penetrate very narrow cracks	
Breeding habits	Similar for both species:	
	• Eggs are laid on rough textiles	
	• Summer hatching time is 4 to 10 days	
	• Winter hatching time is up to 3 weeks	
LARVAE		
Appearance	Similar for both species:	
	• About $1/8$ inch in length	
	• Pearly white in color	
Duration of larval stage	Similar for both species: 50 to 90 days	
Behavior	Spin netting either as flat mat or feeding tube	Spin cases and drag them about
Preferred foods/ environment	Inside folds of clothing: pleats, pockets, collars, cuffs	Hair, feathers, tobacco, spices, hemp, animal skins

seven months is the usual span. Because the larvae are so active, you may not find any on the materials they have damaged.

What Clothes Moths and Carpet Beetles Eat

Clothes moths and carpet beetles consume fabrics and other materials of animal origin. In addition to their well-known taste for wool, clothes moth larvae will also

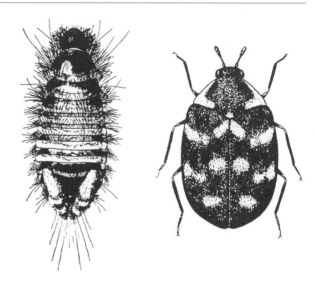

Figure 10.2
Larva and Adult of the
Varied Carpet Beetle
(magnified about
seventeen times)
Source: Carpet Beetles (London,
England: British Museum
[Natural History], Economic
Leaflet No. 8, 1967).

attack paper, straw, cotton, or rayon, using their fibers to spin the cases or mats in which they are transformed into adults.

Carpet beetle larvae prefer napped fabrics, munching the short fibers from the free end to the base fabric in sharply limited spots. They also eat smooth textiles and are especially fond of soiled fabrics. In furs, the grubs lie parallel to the hairs, almost hidden, their heads facing the skin and their bristly rears toward the outer surface. They burrow through packaging material to get at food, clearing the way for other pests. One odd species traces with its chewings the cracks in the floor beneath a carpet. Carpet beetle larvae also eat milk powder, casein, books, seed products, cayenne pepper, dried beans, peas, corn, wheat, and rice.

According to entomologist Walter Ebeling, carpet beetles may even have been found feeding on telephone cable insulation. If they are in an attic, where they feed on dead insects, their cast-off skins can drop through the openings around beams and light fixtures, or through the perforations in acoustical tile. An easygoing homemaker could find that cereal cartons have attracted a carpet beetle gang that will soon invade the rest of the house.

The Hazards of Clothes Moths and Carpet Beetles

Clothes moths and carpet beetles can attack a new sports jacket, a treasured tapestry or Persian rug, a favorite old bathrobe, the horsehair on a violin bow, or even sable-bristled art brushes—they do not discriminate between old and new. Note that in both species, it is the larvae that are the culprits. The adults do no harm.

The following are some of the many materials vulnerable to damage from clothes moths and carpet beetles:

• Woolen or wool-blend items, including blankets, clothing, felt insulation and weather-stripping (most often found in older homes), piano felts, rugs, and upholstery.

• Items made of and/or containing fur, hair, or feathers, including artists' brushes, bows for stringed instruments, down comforters, fur coats, pillows, and upholstery.

• Other substances, including animal skins, cotton, hemp, paper, rayon, spices, straw, and tobacco.

• *Any* material containing traces of beer, fruit juice, fungus spores, gravy, grease, milk, perspiration, or urine.

• And, for carpet beetles specifically, books, casein, cayenne pepper, dried legumes and grains (beans, corn, peas, rice, wheat), milk powder, and seeds and seed products.

How Clothes Moths and Carpet Beetles Make Themselves at Home

Clothes moths thrive in warmer regions, whereas carpet beetles prefer a cooler climate. Fortunately, both insects are less of a nuisance today than they were years ago, since many of our home furnishings and clothes are now made of synthetics, which the pests don't favor. However, we have recently been seeing a return in clothing and home furnishings made of silk, wool, and cotton, and consequently may have to cope with a resurgence of fabric pests.

Any item that uses, gathers, or consists of hair or bird feathers can attract moths. These include:

• Hair and feathers of dead rodents and birds.

• Bird and rodent nests.

• Insect remains.

• Felt insulation and weather-stripping.

• Infrequently changed vacuum cleaner dust bags.

• Piano felts.

• A felt pad under a typewriter.

• Abandoned stuffed toys.

• A rarely used down quilt.

• A pillow with a torn cover.

Any of the listed items can support colonies of moths, whose growing populations migrate in search of more food.

Finding holes in your garments or furniture may be your first tip-off to resident fiber-chewers. Knowing which pest it is can help in dealing with the problem. Carpet beetle holes are spaced rather far apart. The larvae leave their excrement nearby, and it is usually the color of the fabric being eaten. You may also find their cast-off skins, which are pale and rather hairy.

Moth damage is apt to be more localized, since these larvae are more sluggish than those of the carpet beetle. The first reliable sign of clothes moths is usually damaged fabric or the larvae's webbing or cases. (See Figure 10.3.) The occasional moth found wandering around inside your home is most likely another species, possibly a stray from outside or a grain moth.

By contrast, if carpet beetles are clustering on the inside of your windows, this is a signal for firm action. They are reproductive adults, with eggs ready to hatch. You may already have seen their larvae wandering from room to room.

The staging area for an invasion of these insects may be hard to find. In addition to baseboards, radiators, and other places where lint and food crumbs collect, the insects could come from heating and air-conditioning ductwork, wall voids, attics, and subflooring. Nests of birds, wasps, or bees near your house, as well as light-colored flowers, could also be the source.

SHUT THEM OUT

Preventing insect fabric damage is much cheaper than replacing clothing or furniture. If prevention is to work, however, it must be thorough. Regular checks and

Figure 10.3
Caterpillar of a Moth
in Its Case and Side
View of a Caterpillar
Removed From Its Case
(magnified about
four times)

Source: Karl von Frisch, *Ten Little Housemates* (Oxford, England: Pergamon Press, 1960). Reprinted by permission.

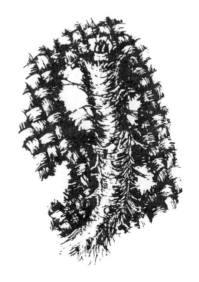

frequent cleaning, especially in attics and basements (often the site of dead rodents or insects and unused wool items), should give you good moth protection. The following precautions for indoors and outdoors should keep both clothes moths and carpet beetles out of your house.

Outdoors

❏ Remove birds' nests (old or new), abandoned spider webs, and wasps' nests hanging under the eaves.

❏ Screen all attic openings to keep out rodents.

❏ Replace any felt insulation around pipes or under the roof with nonorganic material.

❏ Seal all openings around pipes entering the house.

❏ Avoid planting white flowers in your garden. These insects seems to be especially fond of crape myrtle.

❏ When bringing cut flowers inside, check them for feeding carpet beetles.

Indoors

❏ Caulk any spaces between baseboards and walls and floors.

❏ Check your basement and attic, and clean out any dead birds, insects, or rodents.

❏ Clean closets and dresser drawers regularly.

❏ Keep little-used parts of your house free of dust and lint.

❏ Rearrange furniture occasionally to expose carpets to light, since the larvae of these pests prefer dark, secluded corners.

❏ Move heavy furniture once a month so you can vacuum under it.

❏ Vacuum along baseboards and behind radiators. Also vacuum draperies, valances, and upholstered furniture, especially down the backs and sides. If you suspect an infestation, wrap the vacuum bag in a plastic bag and immediately dispose of it in a tightly covered garbage container, preferably in hot sun.

❏ Be aware of little-noticed items like stuffed toys whose owners have long since grown up, mounted animal trophies, and insect collections (keratin pests are a real problem in natural history museums).

❏ Vacuum and wash your pet's bedding frequently to prevent a buildup of sheddings. (As a side benefit, this is also good flea control.)

❑ Watch your spices and cereals, because these also draw carpet beetles.

❑ If you have a piano, vacuum its felts frequently.

❑ If you play the violin or other bowed string instrument, vacuum the case often to protect your bow hairs.

❑ Keep any hog- or sable-bristled art brushes in an airtight box.

❑ Store fabric and yarn leftovers in tightly closed plastic bags. When preparing clothes for out-of-season storage, treat the textile scraps at the same time. (See Proper Handling and Storage of Fabrics on page 179.)

❑ Don't be misled by exaggerated claims regarding the use of cedar linings for closets and dresser drawers. Although regarded as an effective moth barrier by generations of homemakers, cedar vapors give protection only when the smell is strong, and even then, they kill only the smaller larvae. As cedar ages, the vapors weaken, and unless a cedar chest is airtight and the clothes within it are clean, the wool is vulnerable. The wood alone is not enough protection for your woolens. Even fresh cedar can kill young larvae only for a short time, and before too long even this capability is lost. An airtight chest, cedar lined or not, is your best defense against fabric pests.

❑ Remember to sun, air, and brush your valuable, rarely used items from time to time. Also, an attic may or may not be the best place to store your woolens over the summer months. If the temperatures in your area hover in the nineties for days in July and August, the attic heat alone, which could reach 120°F in those conditions, can kill any infestation, since all stages of clothes moths die at 99°F and above. However, an unused attic may contain dead birds and rodents, both of which are attractive to fabric-eaters.

❑ Before the advent of chemical pesticides, many homemakers relied on herbs to hold off insects. While these techniques can still be used, garments so protected must also be properly cleaned and stored. Those legendary housekeepers, the Shakers, used bags of mint and tansy to protect their closets. Another old-fashioned barrier is a mixture of two handfuls each of dried lavender and rosemary with one tablespoon of crushed cloves and small pieces of dried lemon peel, all placed in gauze bags to be kept in dresser drawers. In addition, *The Organic Farmer* recommends all of the following as moth repellents: bay leaves, citrus peel, cloves, lavender, rosemary, santolina, sassafras, southernwood, spearmint, and tansy. Be aware, though, that many of these substances attract pantry pests, so you could be exchanging one insect problem for another. In addition, clothes moths larvae will feed on cayenne pepper, horseradish, ginger, black mustard seed, and hemp—all mistakenly reputed to repel moths.

Proper Handling and Storage of Fabrics

In winter, warm clothing and bedding are safe from fabric pests, if they are in regular use. Just moving the bedding around or wearing your woolens disturbs larvae, which drop off the textile before doing much harm. However, in summer, careful storage is essential.

Blankets, quilts, afghans, and throws should be laundered or dry-cleaned before storing. Dry cleaning destroys all stages of insect life, but does not prevent reinfestation.

HOW TO MAKE YOUR OWN STORAGE FACILITIES

You can make your own storage facilities using such common things as newspaper, adhesive tape, and cardboard boxes to protect your fabrics. The following simple methods may provide all the moth protection you need:

- Put clean, insect-free sweaters, scarves, and blankets into heavy plastic bags or clean cardboard boxes. Seal these with adhesive tape.

- Put clean, insect-free clothes in tightly sealed brown wrapping paper or newspaper packages. If you use newspaper, first wrap the items in an old sheet or towel to prevent ink rub-off.

- Put large items in hanging zippered plastic bags, with their hanger holes securely taped.

HOW TO INSECT-PROOF A STORAGE CLOSET

To insect-proof a storage closet, you will need to do the following.

1. Choose a seldom-used closet.

2. Fill wall and ceiling cracks with putty or Plastic Wood filler.

3. Weather-strip the door with synthetic weather stripping, or seal cracks around the door with strong, wide tape.

HOW TO INSECT-PROOF A TRUNK OR STORAGE CHEST

The following steps will keep insects out of a trunk or other storage chest.

1. Seal any holes or cracks.

2. If the lid does not fit tightly, tape it shut with strong, wide tape.

3. Wrap the whole chest in heavy paper, and tape it shut.

4. Make sure you put only insect-free items into it.

WIPE THEM OUT

A long sunning followed by a brisk brushing, shaking, or beating is one of the most effective measures you can take against fabric pests. To accomplish this, do the following:

❏ Separate the clean items (washed and dried or dry-cleaned, as appropriate) and hang them on a clothesline outdoors for at least eight hours. Hang each piece on its own hanger. As the sun moves around the clothes, any larvae will drop off to get out of the bright light.

❏ While the items are still outdoors, brush each garment vigorously with a good-quality fabric brush; this crushes any eggs, and sweeps any remaining larvae to the ground. Pay close attention to collars (both sides), pockets and flaps, pleats, and seams.

Note: Do not sun furs, as this can cause fading.

If you have no facilities for airing your clothes outside, you have other options:

❏ Put small items—sweaters, scarves, woolen gloves—into your freezer for a few days. This will kill all stages of insect life.

❏ Have the items dry-cleaned or press them with a steam iron. Or you can set your woolens in the oven for one hour at 140°F.

❏ Run the items through a couple of cycles in your dryer set at high heat. Watch that plastic buttons don't start to melt, however. Incidentally, chemical moth crystals can also damage buttons.

❏ For carpet beetles, you can follow the advice of pest expert Hugo Hartnack. In the 1930s, he devised an ingenious trap for carpet beetle larvae. Lay some American cheese on soiled woolens or furs in the corners of an infested room. This will attract the crawlers, which you can then easily catch and drown in hot water. Be sure to pick up and replace the trap regularly.

❏ Smaller household items such as pillows and decorative cushions can be fumigated with dry ice (carbon dioxide, or CO_2), which can be bought at most ice cream stores. To do this, place the items to be treated into a thick-walled plastic bag, seal the bag loosely (this is important, as it allows gas to escape and keeps the bag from bursting), and leave the bag alone until the dry ice has evaporated. When all the dry ice is gone, seal the bag tightly and let it sit for three or four days. Proper fumigation gives quick and satisfactory control, killing all stages of fabric pests. However, it does not prevent future infestations.

Caution: When fumigating with dry ice, work in a well-ventilated area, as CO_2 replaces oxygen in the air. Also, do not touch dry ice with your bare hands. Instead, wear heavy gloves and/or use tongs to touch the dry ice.

SILVERFISH AND FIREBRATS

Silverfish and firebrats are sometimes called fish moths, tassel tails, or fringe tails. Probably the oldest insect species on earth, their ancestors lived about 400 million years ago. They can move quickly, even sideways. Only their scaly, silvery appearance gives silverfish their name, for they do not swim. (See Figure 10.4.) Firebrats look like silverfish but are slightly larger.

Although silverfish prefer cool, damp places, particularly basements and laundry rooms, they can survive in any room of the house. They have been found in Florida attics with temperatures as high as 130°F and humidity below 15 percent. One species is often seen in large numbers in houses with wood roof shingles. Firebrats seek very warm, moist areas, like those around ovens and fireplaces. They lurk in attics in summer, and near furnaces in winter months.

What Silverfish and Firebrats Eat

These pests eat the sizing on paper and chew ragged holes in the paper itself. Although they avoid wool, they do eat linen, cotton, silk, clothing starch, rayon, paste, glue (such as that used in book bindings), cereals, and dead insects.

The Hazards of Silverfish and Firebrats

Silverfish and firebrats, which eat carbohydrates, can ruin your library or family papers. They also attack textiles of vegetable origin, like cotton and linen, and seem especially fond of rayon.

How Silverfish and Firebrats Make Themselves at Home

These insects invade our homes via several routes, among them the following:

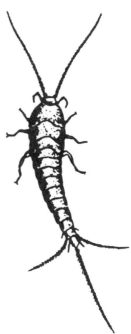

Figure 10.4
The Silverfish

Source: Insect and Rodent Control: Repairs and Utilities (War Department Technical Manual, TM-5-632, October 1945).

- You may carry them in with secondhand books, cardboard boxes, or old papers.

Treating Damaged Items

If the prevention advice outlined in this chapter comes too late for some of your cherished woolens, you can have the damaged pieces that are worth repairing cleaned and the holes rewoven. It can be expensive, but, properly done, the mend is invisible. Badly infested garments should be destroyed, however, either by burning or by setting them in a plastic bag in a closed garbage container and baking the whole thing in the sun.

• If your home is new, they may have come with the wallboard and green lumber, and found a feast of wood shavings, sawdust, and wallpaper paste.

• These long-lived nocturnal insects could be making their way from inside the flowers near your home's foundation.

Once indoors, these pests hide and lay their eggs in wall cracks and behind baseboards and loosened wallpaper. Their eggs can take up to three years to pass through all the stages of insect development. An infestation develops slowly and if you find them occasionally in your home, they have probably been there a long time. If you find one in the bathtub or sink, it has fallen in accidentally and cannot climb up the slippery sides to escape.

Apart from seeing the insects themselves, the first signs of these pests are the damage they cause to paper goods, books, or fabrics. To determine if these are the culprits damaging your fabrics or other items, place a card coated with flour paste near the cloth. Feeding marks on the card will tell whether or not silverfish, who like flour, are doing their dirty work.

SHUT THEM OUT

Preventing and controlling silverfish and firebrats is much less traumatic than confronting their ravages on irreplaceable books and documents or expensive clothing. Prevention begins outside and then moves indoors, with the following precautions.

Outdoors

❏ Close off any holes around pipes entering walls.

❏ Relocate flowers growing around foundations to beds that are some distance from the structure, because these insects like the mulch in flower beds.

❏ Avoid using bright white electric lights outdoors; use yellow lights instead, and then only when really necessary.

Indoors

❏ If you collect old books, check all new acquisitions carefully before setting them on your shelves.

❏ Clean out bookcases periodically, shaking out the books. If you're a book-lover like me, this is a chore that can kill much of a day, but then when you reach for a treasured volume you will be pretty certain you won't find any chewed pages or binding.

❏ Repair any plumbing leaks.

❏ Check any lined draperies you have; silverfish often hide between the lining and the outer fabric.

❏ If you live in an apartment, check the basement incinerator from time to time. If it is poorly maintained, this can provide a perpetual free meal for silverfish and firebrats, as well as for other pests.

WIPE THEM OUT

Thorough cockroach control invariably eliminates silverfish and firebrats.

Like cockroaches, these insects can be poisoned with *technical* boric acid. The powder should be blown with a bulb duster into any baseboard crevices and around door and window frames. Control may take several weeks. (See page 47 for a complete discussion of buying and using boric acid.)

One unique trap for silverfish is easy to construct. A small glass jar, its outside (and only the outside) covered with masking or adhesive tape, is all you need. The insect crawls up the tape, falls into the jar, and can't climb up the slippery inside walls. These traps should be set in the corners of the infested bookcase or pantry at the corner of a baseboard, or at any intersection of wall and floor. Some authorities advise baiting the jar with a bit of wheat flour or chipped beef; others say this isn't necessary.

As you can imagine, silverfish and carpet beetles are a major concern to librarians. How do the custodians of large collections protect their holdings? Except for massive infestations, the librarians at the University of California, San Diego, have stopped using chemical pesticides. They say, "When material can be removed or is a new acquisition, freezing is the standard method of pest control. The procedure is safe, lower in cost than fumigation, and effective." Some museums protect their

Why Not Chemical Insecticides?

The chemicals sold to protect furniture and clothing against insect damage can expose you and your family to unnecessary health risks. Paradichlorbenzene (PDB) and naphthalene, two chemicals widely sold to control clothes moths, are both poisonous to humans. Benzene, the basic ingredient in PDB, is a known carcinogen; naphthalene can cause violent blood disorders in sensitive people, especially those with dark skin. More than 5,000 people, including 4,000 children under the age of six, are poisoned every year by either eating these toxins or breathing them in. More than 1,000 require treatment at a health-care facility.

What about camphor? Used for generations to protect moth-susceptible fibers, camphor is now recognized as toxic to humans. It can cause serious nerve disease, and people who have accidentally eaten camphor balls have died. Camphorated oil, an old remedy for respiratory illnesses, has killed some children when it was rubbed on their chests. Curious children often pop antimoth nuggets into their mouths, thinking that they are candy.

If your winter clothes, generously sprinkled with chemical antimoth crystals, are stored in a bedroom, sleepers are breathing in the fumes. And if you don't have your sprayed garments dry-cleaned before wearing them again, you could be absorbing toxic chemicals through your skin. Other protective measures work just as well as these chemicals, and with much less risk.

holdings by keeping them in sealed containers constantly flooded with nitrogen gas, suffocating the insect by eliminating the oxygen necessary to life.

CRICKETS

You can forget about Pinocchio and his lovable six-legged sidekick, Jiminy. Crickets, which are sometimes mistaken for cockroaches, help turn a summer's evening into a fresh-air concert, but if they should invade your home, *watch out*. Their chirping (actually a mating call), which is so pleasant outside, can drive you wild when you're trying to sleep, and they can cause damage. Since the male's song attracts the females, although you may be hearing only one chirper, he is probably attended by a sizable and voracious harem.

Chinese and Japanese cricket fanciers keep them as pets for good luck, often in ornate cages. These imaginative people pit two aggressive males against each other and bet as much as ninety dollars on the outcome of a fight.

The Hazards of Crickets

Although crickets are not found indoors as often as the other pests discussed in this chapter, once inside, they can be even more destructive. They will feed on and can destroy almost any fabric, chewing large holes in anything they attack. They are especially fond of silks and woolens, and can also damage nylon, cotton, carpeting, even leather.

How Crickets Make Themselves at Home

Where do crickets come from? Poorly managed landfills and dumps are frequent sources of cricket infestations—as are litter-strewn gardens. Moisture also encourages some species, particularly the cave, or camel, cricket. Crickets can be carried in on firewood, or, when their natural food supply of grass dries up, they may invade the home, especially in the fall.

If one or two hopping about in your kitchen disturb you, be grateful that you weren't a resident of Coalinga, California, a few years back, when a massive invasion of crickets knocked out the power plant and turned the streets into one big pool of slime, which was cleaned up only when skip loaders were brought in to clean up the mess.

SHUT THEM OUT

As with most pest problems, prevention is the simplest and safest form of control.

Outdoors

❏ Because this relative of the equally musical katydid hides under leaves and other organic litter during the day, keeping your garden free of these havens will discourage crickets, as will pulling up the large areas of broad-leafed ground covers, which are often used instead of grass.

❏ Be aware of tilling or discing for weed control in nearby fields. Weed destruction there may force the crickets to leave, searching for food, and they may migrate to your property.

Indoors

Beyond garden housekeeping, controls similar to those used for cockroaches should keep crickets at bay.

❏ Make sure any cracks in your house's exterior walls and around the foundations are tightly caulked. Doors and windows should fit snugly, and all vents should be adequately screened. Any wall cracks should be repaired, and base-boards should fit tightly against walls and floors. A caulking gun, patching plaster, a screwdriver to tighten moldings and cabinet frames, and a good vacuum cleaner may be your most useful tools in your pest control efforts.

❏ Like silverfish, crickets are drawn to light, so use your outdoor lighting sparingly, and use yellow rather than white bulbs.

WIPE THEM OUT

❏ If, despite your best efforts at prevention, crickets establish themselves in your home—which is most likely to happen in early fall—vacuum them up, empty the vacuum bag into a plastic bag, place the bag in a garbage barrel with a close-fitting lid, and set the barrel in the direct sunlight. Then blow a desiccating dust like silica gel or diatomaceous earth (the kind used for insect control, *not* the kind used for swimming pools) into inaccessible wall voids. If you use either of these substances, be sure to wear protective gear; although nontoxic, they are very irritating.

❏ Some householders use duct tape to trap crickets. They lay out strips of the tape, adhesive side up, near where the chirpers are hiding. The adhesive evidently attracts the insects, who stick to the tape, and can then be easily destroyed.

With all these small marauders waiting to savage our clothes, home furnishings, and records, it's a wonder we all aren't threadbare. The truth is that reasonably careful housekeeping will hold damage to levels we can tolerate—without chemical insecticides.

11. Termites and Carpenter Ants— The Hidden Vandals

In the seventeenth century, La Rochelle, the Huguenot port from which the French explorers Samuel de Champlain and Jacques Cartier sailed for the New World, fell to the cannons of Cardinal Richelieu. Rebuilt, La Rochelle almost fell again two centuries later—this time to an insect as destructive as any cannon. When tropical termites invaded the town in the nineteenth century, whole streets were undermined. The arsenal and chief government building had to be shored up, and all the town records were chewed to a pulp.

During World War II, crates containing military supplies, temporarily stacked on the ground in New Guinea, disintegrated when they were moved a few weeks later because they had been invaded by subterranean termites.

Although just as destructive in some regions as termites, carpenter ants do not eat wood. Instead, they tunnel into it to create their nesting sites. In this chapter we will consider both.

TERMITES

Like their cockroach ancestors, termites speed the transformation of dead forest trees into soil nutrients, a process that might otherwise take hundreds of years. In earlier times, these "white ants" eased American pioneers' heavy task of settling and cultivating new land. After trees were felled, the insects gradually cleared away the stumps, and in the process enriched the soil for the farming that followed.

In Sri Lanka, three-fourths of the claylike land would be unfit for agriculture without the teeming termites loosening the soil so that rain can penetrate it. Tropical termite mounds, complex structures that are often as much as twenty or thirty feet high, are made of termite excretions that harden into a material so tough that humans can use an abandoned hill as a baking oven or metal smelter. Crushed, the substance makes a resilient, long-lasting surface for a tennis court. In the Sahara, termites are critical in helping reclaim soil damaged by drying heat and winds and overgrazing by livestock.

Unless you live in a temperate region, you probably have never seen a termite. If you did, you might have confused it with a winged ant. (See Figure 11.1)

Types of Termites

Scientists have divided termites into two major groups; those that live in the earth (subterranean termites), and those that live above it (drywood termites). Subterranean species cause the most damage.

Buildings in warm, moist regions are particularly vulnerable to subterranean termites. Along the Gulf and Pacific coasts, the drywood termite, which needs no soil, causes considerable havoc. The only state that is considered free of termites is Alaska.

Subterranean Termites

Subterranean termites live in colonies, which can number from hundreds of thousands to several million individuals. They nest in the earth, sometimes thirty feet down, and may feed on buried wood or wood above ground. To reach wood above the surface, termites tunnel up, building flattened shelter tubes to protect themselves from predators when they must emerge from the soil. When not feeding, termites return underground to replenish their body moisture. The reproductives members of the colonies, called "queens" and "kings," can live for decades; workers, for several years. The faithful king stays with his queen for life, constantly fertilizing her.

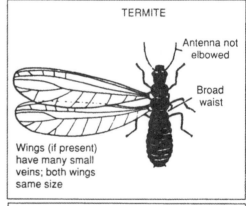

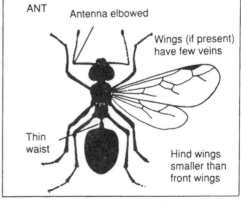

Figure 11.1
Major Differences Between
Winged Ants and Termites

Source: Termites and Other Wood Investing Insects (Division of Agricultural Sciences, University of California, Leaflet 2532, 1981).

Within the last twenty years, a different species of subterranean termite, probably originating somewhere in Asia and called the Formosan termite, has crossed the ocean to Texas and Florida. More aggressive in its tunneling than our native troublemakers, the Formosan termite is also more resistant to the soil poisons that control other species. New nonchemical techniques that infect other subterranean termites with a fatal disease are ineffective against the Formosan. Members of this species will bite off the legs of a sick comrade so that it cannot spread contagion throughout the colony.

Figure 11.2 maps the areas of relative subterranean termite activity in the United States.

Drywood Termites

Drywood termites, whose colonies are relatively small, nest in the wood they eat and never need ground contact. These insects can go without moisture for long stretches, so they are often carried in furniture and crates from one area to another. Although they reproduce slowly (not more than fifty to a hundred young a year), drywood infestations spread easily.

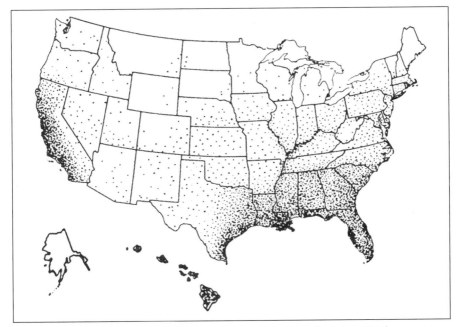

Figure 11.2 Relative Hazard of Subterranean Termite Infestation in the United States

Source: Subterranean Termites, Their Prevention and Control in Buildings (Washington, DC: United States Department of Agriculture Home and Garden Bulletin 64, 1979), adapted.

What Termites Eat

A termite's diet is composed of wood and wood products like paper, wood pulp, and fiberboard. Whether that wood takes the form of a fallen log in a forest or a home in an upscale neighborhood makes no difference to this insect.

The Hazards of Termites

Despite their benefits in nature, termites become expensive pests where we humans use wood to make our homes. In the United States alone, hundreds of thousands of homes are treated every year for termites. The total damage to houses annually in this country is about $5 billion, roughly equal to that caused by fires.

How Termites Make Themselves at Home

Termites are an ever-present danger to homeowners, even those with brick houses. A crack in mortar or cement only a thirty-second of an inch wide will let these pests crawl under the brick to the wood. They will follow an ooze of water to moist wood sometimes ten feet away. One species was found on the fifth floor of a concrete fireproof building in San Francisco.

Where subterranean termites are not abundant, a house that meets modern building codes could remain sound for many years. On the other hand, if a house is built over or near an existing colony, winged reproductives can show up after only three or four years, and structural damage might be evident some years after that. Along the Gulf Coast, some of the ground is continuously infested with subterranean termites, and a house may be attacked as soon as it is built. Drywood termites, common in California, the Gulf states, and the islands of the Caribbean, can also invade immediately. Incidentally, houses with attached garages whose doors are kept open much of the time have a greater chance of invasion by drywood termites, which may be blown in from dead trees and telephone poles.

Homes over thirty-five years old are those most likely to have termite damage. In some regions, every home on a block in an older neighborhood may show signs of termite invasion.

Even without the Asian immigrant, the termite problem seems to be getting worse. Aging buildings and patios, breezeways, and attached garages—all increasingly popular home features—open the way for termite invasions. Slab construction, now widespread, makes control difficult because with this method, wood is closer to the ground and concrete tends to crack, opening the way for underground wood pests. As our virgin forests with their mature trees disappear, more homes are built with sapwood from younger trees, which is more susceptible to termites. Some types of wood, however, are more termite resistant than others, among them

Arizona cypress, black walnut, and several varieties of oak. Finally, central heating encourages the insects' activity year-round.

Because these pests don't like to be shaken, objects such as railroad ties and musical instruments that vibrate frequently are relatively safe from assault. If you need a good reason to play your piano, this is it.

Termites do their dirty work out of sight, deep within our walls or furniture. By the time some of their destruction is visible, they are well established and difficult to eradicate. However, they do send out some early signs that can warn you of a developing problem.

• *Swarming of the reproductives,* the potential "queens" and their consorts, as they fly out of the colony is one sure clue to all termite species. This usually happens in fall and spring. (Ants also have reproductive swarms.) After a brief mating flight, the sexually mature insects shed their wings (as do ants) before burrowing into wood to begin laying eggs. A mass of tiny wings scattered on the ground is a sign that termites are thriving in the area. At this stage, we humans have many helpers destroying termites: Spiders, ants, lizards, and birds all gorge themselves on the now-wingless, helpless pests. Rare indeed is the termite couple that survives. The predators know what they're doing. Some of us humans agree with them. In southern Africa, fried termites are sold as snack foods in many markets, or served with tomato and porridge to make a satisfying lunch.

• *Shelter tubes,* built of particles of earth and wood cemented together with a glue-like secretion, are a sign of subterranean termites. Besides protecting the foraging insects, the tubes provide an exit for the swarming reproductives. The tubes may be clinging to a foundation or hanging from a joist or girder.

• *Wood flooring with dark or blistered areas that are easily crushed* with a screwdriver or kitchen knife signifies subterranean termites. To verify your suspicions, rap on the suspected wood. If you hear an answering tapping, rather like the ticking of a watch, the subterranean soldiers are alerting the colony to danger by banging their very hard heads against the gallery walls.

• *Fecal pellets* are the calling card of drywood species. Roughly oval and six-sided, about the size of a sesame seed, the pellets are generally seen near the pests' "kick-out" holes. Drywood termites are excellent housekeepers and sweep all the waste out of their galleries. Wooden sheathing that gives way easily when pressed can also tell you you have drywood termites. We had to have part of an exterior wall replaced when the drywood termites infesting a neighbor's fence moved to an outside wall of our house. They never went to the fence owner's house, and a termite inspection revealed no signs of infestation in the neighbor's walls. There's no justice in the insect world!

SHUT THEM OUT

Knowing something about how to prevent termites will lessen your chances of needing professional control. Termite prevention begins with sound construction.

Preventing Subterranean Termites During Construction

Following standard building codes carefully is the best protection against subterranean species. If your home is under construction, be aware of the following steps to forestall the problem of subterranean termites:

❏ Make the structure as dry as possible:

* The lot should be graded to let water drain rapidly away from the house and its adjacent structures.

* Gutters, downspouts, and foundation drains should all guide water clear of the building.

* Doors, windows, roof valleys, and chimneys need to be adequately flashed.

* Attics and crawl spaces should be well vented to prevent accumulations of moisture.

* Walls should be fitted with vapor barriers.

* In very moist regions, the soil under a slab-based house should be shielded with a heavy polyethylene sheet before concrete is poured.

* Make sure refrigeration and evaporative coolers drain well so the wood around them stays dry.

❏ Apply soil poisons. Slab construction, once thought to be effective termite proofing, may actually open the way for an infestation. Temperature changes, earthquakes, sonic booms, and natural earth shifts all can crack concrete, giving termites easy access. One of the surest defenses is soil poison applied before building begins. It's a relatively safe procedure, according to the U.S. Department of Agriculture. Tests have shown that oil-soluble insecticides stay in place, moving neither sideways nor down toward the water table, and can protect a structure for twenty-five years. As described on page 196, a sand barrier will also protect a home.

❏ Use a poured concrete foundation. Poured concrete foundations, which are easily inspected, are the most termite-resistant form of residential construction. In addition, a four-inch-high reinforced-concrete cap between the foundation and the subflooring will expose any termite tubes. Metal termite shields, once thought more effective than the concrete cap, are difficult to install properly, and consequently don't work as well.

❏ Handle wood properly. Since termites must eat wood or wood products, the critical factors in building are the following:

- Treat wood with repellents or insecticides before it is incorporated into the structure.

- Permit no wood to touch soil at any point.

- If possible, use wood known to be termite resistant; among these are Arizona cypress, black walnut, and some varieties of oak. The Uniform Building Code, which is followed in about twenty states, requires the use of treated wood wherever wood is going to be in contact with masonry or concrete. Since subterranean termites have to tube across treated members to reach the untreated ones, an infestation is easily detected. Standard codes also require that no wood penetrate concrete to the soil below; that there be adequate clearance between structural wood and the soil; and that a crawl space large enough for thorough termite inspection be provided.

- When the house is finished, be meticulous in clearing away all wood debris, especially from earth-filled porches, steps, and patios. Wood scraps used for fill are responsible for over half the infestations of subterranean termites. Yet, "I've never heard of a building inspector digging through backfill to look for wood debris," says Dr. George Rambo, formerly Director of Technical Operations for the National Pest Management Association and now president of his own consulting company. If your home is under construction, make sure that all backfill is free of wood debris, even if you have to take a shovel and dig through it yourself.

Figure 11.3 shows twenty points of careless construction and home maintenance that can open the way for subterranean termites. Incidentally, conditions that favor termites also favor wood fungi, so it is vital to replace any wood that shows signs of rot as soon as possible.

Preventing Drywood Termites During Construction

If yours is an area with drywood termites, before the house is decorated, consider having a pest control operator drill holes in the plaster or wallboard and blow in a silica aerogel dust. The operator can also apply the silica aerogel dust through the attic access hole, using one pound of dust for every 1,000 square feet. In both cases, the dust—which is nontoxic—will remain in place indefinitely, killing any insect that crawls over it. Some building companies spray lumber with liquid solutions of boric acid during framing, a step that can be very effective in preventing drywood infestations.

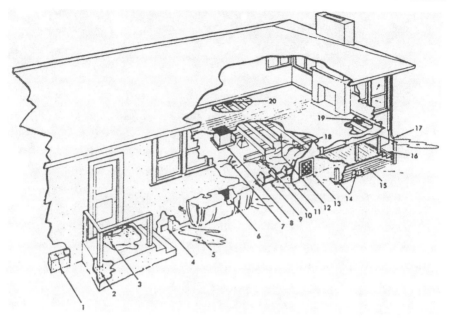

Figure 11.3 Faulty Home Construction and Maintenance Promoting Subterranean Termite Infestation

Source: Walter Ebeling, *Urban Entomology* (Berkeley, CA: Division of Agricultural Sciences, University of California, 1978 [rev]).

Preventing Termites in Homes Already Built

❏ A professional termite inspection is your best insurance against infestation in a house already built. Such an inspection is a difficult, often dirty and uncomfortable, time-consuming job requiring agility and expertise. The inspector must probe all your home's hidden recesses, crawling through dark, cramped places to examine drains, floor furnaces, and attics. The inspector puts his or her own integrity and the company's liability on the line with every inspection. Be prepared to pay a reasonable fee for the service.

By law you should receive a formal, written report of the findings, including a clearly labeled diagram of the house indicating any infestations, all termite-favoring conditions, and areas that could not be inspected. If you feel the inspection wasn't thorough enough, arrange for a second. Figure 11.4 is a standard structural pest control report form, and Figure 11.5 is a sample diagram of an inspected house.

Your state may also require a second report to be filed with the agency overseeing the pest control industry. This report indicates what remedial work has and has not been completed. In California, for a nominal fee, the Structural Pest Con-

STANDARD STRUCTURAL PEST CONTROL INSPECTION REPORT
(WOOD-DESTROYING PESTS OR ORGANISMS)

This is an inspection report only · not a Notice of Completion.

ADDRESS OF PROPERTY INSPECTED	BLDG NO	STREET	CITY	DATE OF INSPECTION
			CO CODE	

FIRM NAME AND ADDRESS			

Affix stomp here on Board copy only

↓ A LICENSED PEST CONTROL ↓
↓ OPERATOR IS AN EXPERT IN ↓
HIS FIELD. ANY QUESTIONS
RELATIVE TO THIS REPORT
SHOULD BE REFERRED TO HIM.

FIRM LICENSE NO	CO REPORT NO (if any)	STAMP NO

Inspection Ordered by (Name and Address) _____
Report Sent to (Name and Address) _____
Owner's Name and Address _____
Name and Address of a Party in Interest _____

INSPECTED BY _____ LICENSE NO _____ Original Report ☐ Supplemental Report ☐ Number of Pages _____

YES	CODE	SEE DIAGRAM BELOW	YES	CODE	SEE DIAGRAM BELOW	YES	CODE	SEE DIAGRAM BELOW	YES	CODE	SEE DIAGRAM BELOW
		S-Subterranean Termites			B-Beetles-Other Wood Pests			Z-Dampwood Termites			EM-Excessive Moisture Condition
		K-Dry-Wood Termites			FG-Faulty Grade Levels			SL-Shower Leaks			IA-Inaccessible Areas
		F-Fungus or Dry Rot			EC-Earth-wood Contacts			CD-Cellulose Debris			FI-Further Inspection Recom

1. SUBSTRUCTURE AREA (soil conditions, accessibility, etc.)
2. Was Stall Shower water tested?   Did floor coverings indicate leaks?
3. FOUNDATIONS (Type, Relation to Grade, etc.)
4. PORCHES ... STEPS ... PATIOS
5. VENTILATION (Amount, Relation to Grade, etc.)
6. ABUTMENTS ... Stucco walls, columns, arches, etc.
7. ATTIC SPACES (accessibility, insulation, etc.)
8. GARAGES (Type, accessibility, etc.)
9. OTHER

DIAGRAM AND EXPLANATION OF FINDINGS (This report is limited to structure or structures shown on diagram.)

General Description _____

Signature _____

YOU ARE ENTITLED TO OBTAIN COPIES OF ALL REPORTS AND COMPLETION NOTICES ON THIS PROPERTY FILED WITH THE BOARD DURING THE PRECEDING TWO YEARS UPON PAYMENT OF A $2.00 SEARCH FEE TO STRUCTURAL PEST CONTROL BOARD, 1430 HOWE AVENUE, SACRAMENTO, CA 95825.

Figure 11.4 Standard Structural Pest Control Report

Source: So, You've Just Had a Structural Pest Control Inspection (Division of Agricultural Sciences, University of California, Leaflet 2999, 1980).

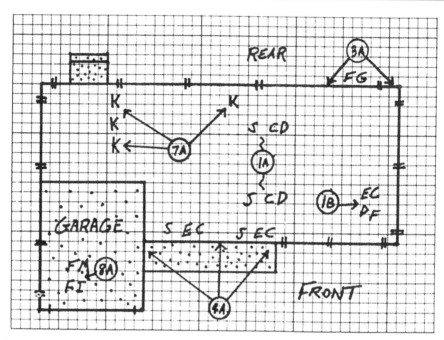

Figure 11.5 Termite Inspector's Diagram of a House

Source: So, You've Just Had a Structural Pest Control Inspection (Division of Agricultural Sciences, University of California, Leaflet 2999, 1980).

trol Board will provide you with a copy of every report filed within the past two years for any property that you are thinking of buying. It would be prudent to track down such reports no matter what state you live in.

❑ Use sand to create an effective termite barrier. A method that is harmless to humans, yet promising in effectiveness against subterranean termites, is the laying down of a sand barrier to stop the insects' upward tunneling. In sand composed of the proper size particles, the tunnels collapse. In 1989, the city of Honolulu incorporated the use of antitermite sand barriers into its building code.

Entomologist Walter Ebeling first conceived of the sand-barrier idea in 1957, but, at the time, the process seemed to be only of academic interest. The lure of the quick chemical "fix" was too strong. Thirty years later, Professor Ebeling, along with the late earth scientist Charles F. Forbes, devised and patented a practical method for applying sand to block termites.

To shield an entire slab-based house from subterranean termite invasions, the sand barrier must be laid down before construction. However, termite infestations originating *outside* the house (as opposed to underneath it) can be combated by applying the sand barrier against the outer surface of the slab foundation. A house built on a conventional footing can be protected with sand after construction. Sand with particles in the range of 10- to 16-mesh replaces soil around a building's foundation and sometimes even in the crawl space. The insects can't tunnel through the coarse sand and so are effectively blocked from invading. This technique, using volcanic sand, is widely used in Australia and Hawaii and has been tried with some success in the continental United States. Stainless-steel screening may soon be on the market for the same purpose.

Checklist for Keeping Your Home Termite-Free

❑ Keep all plumbing in good repair.

❑ Make sure your shower pan is free of leaks.

❑ See that basement air vents are fully exposed, not overgrown with shrubbery.

❑ When watering, avoid sprinkling stucco or wood siding

❑ Fill any cracks in masonry or concrete with cement grout, roofing-grade coal-tar patch, or rubberoid bituminous sealers, which should be available at better home and garden centers or hardware stores.

❑ Keep gutters and downspouts in good repair.

❑ Finally, if *common* ants are hard at work around your foundation, leave them alone—and be grateful. They are nature's termite inspectors, always on the job. Termites have many predators—birds, lizards, beetles—but their greatest foe is the ant. War between the two is eternal.

WIPE THEM OUT

Although this book focuses on controlling pests without poisoning ourselves, termite control is one area in which chemicals may be necessary. Applying these pesticides is a job for a well-trained professional.

Chlordane was the most widely used poison for controlling subterranean termites until the mid-1980s. It was also used in agriculture. However, it is strongly suspected of causing cancer and other serious diseases in humans. It is long-lasting—which is a big plus in termite control—but it accumulates in the body's fatty tissues.

In Los Angeles, a beautiful older home that had been treated for subterranean termites in the late 1970s became a poisonous trap for its owners. For a year after chlordane was sprayed into the structure's crawl space, the family members suffered repeated headaches and nausea. In an effort to make the house livable, the owners had the entire heating and ventilating system reworked, but no one—not the U.S. Environmental Protection Agency (EPA), the county health authorities, or the National Academy of Sciences—could define for them a safe level of exposure to chlordane.

In 1983, a young Long Island family had to see their contaminated home bull-dozed because of incompetent termite treatment.

Admittedly, these are extreme cases. The use of chlordane is now prohibited in the United States, but it is still produced here for export to other countries. More-over, it remains a problem because it is so stable and persistent. The EPA thinks that as many as 80 million homes in this country may be contaminated with chlor-dane residues, something to consider if you are thinking of buying an older home.

Chlorpyrifos (marketed under the trade names Dursban and Lorsban) is another hazardous substance that has been widely used against termites. Its long-term effects are weaker than those of chlordane, but, in the short run, it is more toxic. The EPA banned the use of chlorpyrifos in existing homes in 2000, and it is now illegal to use it on new construction as well. Some pest control operators may still be using it, however. Moreover, it is not necessarily always applied by some-one certified by the state or federal government as competent to do so. And appli-cators do not always follow the directions on the chemical's label.

New, less toxic termiticides have been developed to replace these dangerous long-lasting substances, but their effectiveness has been much weaker, with failure rates as high as 30 percent in the first year.

As the public becomes increasingly aware of the serious health risks of using termiticides in homes, science and industry are joining forces to develop and mar-ket benign alternatives to control this stubborn pest. Among the methods promis-ing to be effective are microorganisms that destroy subterranean colonies, and sand barriers that prevent underground tunneling. Heat and electrocution may be useful against drywood termites, as may extreme cold. However, none of the newer controls is nearly so long-lasting as chlordane, so houses may need to be inspected and treated for wood pests as often as every five years.

Treating for Subterranean Termites

❏ Consult a qualified pest control operator. This is a job for a professional, and the appropriate technique for treating underground termites depends on the struc-ture's foundation. Slab construction may require drilling close to the site of the infestation and injecting a termiticide into the soil beneath the house. In another approach, insecticide-carrying pipes are introduced horizontally across the slab to saturate the soil. A house with a crawl space is the easiest to treat. Applicators dig a shallow trench around the foundation, pour the chemical into the trench, and replace the soil. Or they may insert a nozzle about four to six inches into the soil and release the liquid termiticide. (See pages 200 through 203 for a full discussion of how to choose a pest control operator.)

There are also a number of nonchemical approaches that may be tried. In one nonchemical approach, a biological substance (a nematode, a microsized worm) is introduced into the soil. Its microorganisms home in on the pests, which sicken and die. Other termites in the colony get the disease when they consume the victims (termites are cannibals). This procedure, too, should be done only by a professional.

Treating for Drywood Termites

Drywood termites must be handled very differently from those that invade from the soil. If the infestation is clearly localized, "drill and treat" is the most common method. Holes are drilled into the insects' galleries, and chemical poisons injected into them. Today the pesticide imidacloprid, marketed as Premise, is often used for this method. It is highly toxic to insects and aquatic life, but of very little danger to the warm-blooded. In a weaker form it is used as flea control on cats and dogs.

Because it is difficult to determine the extent of a drywood infestation, pest control operators prefer to use fumigation against these pests. (See page 201.) After drill and treat, the question always remains, "Did they miss some?" After fumigation, the answer is generally no. However, even fumigation sometimes fails to kill all the termites.

If the termites are in only a single piece of furniture, you can have the piece fumigated away from your home. Some pest control companies and some county agricultural offices provide this service.

As with subterranean termites, there are a number of nonchemical approaches to dealing with drywood termites. (Please note, however, that if you are planning to sell your house in the near future, you should check with a local realtor to find out whether mortgage lenders in your state accept any of these alternative methods of termite control. Financial institutions in California do, but banks in other states may not.) Among available alternative methods are the following:

• *Electrocution.* A device called the Electrogun sends a powerful current of electricity through the termite galleries, electrocuting the pests or so disrupting their life systems that they soon die. This method has been used for a number of years with good results, and is very safe. We have had it done several times, and apart from being warned not to stand on any metal during the process (the device runs on 90,000 volts of electricity) the only inconvenience we felt was the loss of a telephone, which the pest controllers replaced.

• *Heat.* Most insects, including termites, can tolerate only a narrow range of temperatures. Professors Ebeling and Forbes as well as others, seeking effective nontoxic control of drywood termites, found that the pests died within thirty minutes of exposure to temperature of 120°F. These practical academics then devised a technology for delivering such heat to the interiors of a house's wooden members, where termites build their galleries. The heat can be applied to individual rooms or entire buildings.

To induce such a high temperature in the termite galleries, air that is hotter than 120°F—up to 166°F—must be blown into the building with high-velocity propane heaters. Sensitive household items that may be damaged by such temperatures are removed from the building first.

This heat treatment is now available from a number of pest control operators who have been trained in its use. You would be wise to check around for a company equipped to offer this service. Heat treatment, which may require whole-house tenting (depending on the extent of the infestation), should greatly reduce the need for chemical fumigation.

• *Cold.* Some pest control operators think that extreme cold produced by liquid nitrogen holds promise for killing drywood termites in their galleries. However, this method is suitable only for spot treatments, as it requires the drilling of holes into walls to reach studs and joists, causes walls to "sweat," and may turn out to be more costly than heat or electrocution.

The use of both liquid nitrogen and the Electrogun depends on pinpointing the sites of termite galleries, which are often inaccessible to human inspectors. Here is a situation in which a dog can truly prove to be your best friend. Specially trained beagles, with their keen hearing and sense of smell, can squeeze into narrow crawl spaces, spot a hidden colony, and alert a pest control operator to active infestations. However, although they are a great public-relations ploy, dogs do make mistakes and sometimes will indicate an infestation months after all termite activity has stopped.

HOW TO FIND A COMPETENT PEST CONTROL COMPANY

Homes have been made unlivable and families have become ill because of poor termite-control work. Close, competent supervision is the key to safe, effective commercial pest control.

How do you go about finding a pest control company you can trust? Very carefully:

❏ A good start is a talk with a well-established local realtor. He or she will probably know qualified pest control operators in your area.

❏ Get a recommendation from a friend who has had a termite problem and been satisfied with the company that handled it.

❏ Check the yellow pages of the telephone directory. The ads there, ranging from the catchy to the businesslike, may present a confusing choice. My own preference is to call companies that avoid the bug cartoons and solicit my business in a straightforward way.

❏ Contact at least three firms. It may or may not be to your advantage to talk only with companies that have been around a long time. The outfit that contaminated the Los Angeles home mentioned earlier had been in business for fifty years.

❏ Ask the right questions. Begin to ask questions on the telephone. Suggested questions include the following:

• Is the salesperson who will come to the house licensed to make recommendations on termite control? If the answer is no, call another company.

• How much of the firm's business is termite control? In areas where termites abound, it should make up about 30 percent of the average company's business, if they are licensed for this category.

• How will the applicators be supervised?

• Are the applicators literate?

• How are the applicators trained? Do they and their supervisors receive continuing education in pest control?

• What insurance does the company carry? General liability and worker's compensation should be the minimum. Coverage for errors and omissions would also ensure that the company could reimburse you if the crew fouls up.

• Is the firm familiar with, and does it use, advanced, nonchemical controls where possible?

Note: If your state supervises pest control closely—California, Maryland, Florida, and North Carolina are among the best—your questions may not need to be so detailed as these, but if in doubt, it is never a bad idea to ask.

❏ Learn about fumigation. Fumigation, the usual treatment for dispersed infestations of drywood termites, requires a special set of questions. *Safety must be the absolute prime consideration in fumigation.* The gases used are odorless, colorless, tasteless—and deadly. The company's representative should be crystal clear about how you and your property will be protected. Don't be hesitant about asking any or all of the following questions:

• How will foods be protected?

- What should be done with gas-sensitive plants, both inside and outside the house?

- Will photographic equipment be affected?

- What will happen to valuable art works like oil paintings?

- Will rubber and leather items be affected? If methyl bromide (a fumigant once used but technically supposed to be phased out) is applied, then any rubber with which it comes into contact, including that in your carpet padding, will develop a persistent, unpleasant odor. Leather also may be affected. Sulfuryl fluoride, known commercially as Vican, commonly used today, affects neither. However, it may cause some metals to corrode.

- Will weather affect the fumigation? The colder the day, the longer it takes for the gas to kill the pests and decompose. If the temperature is expected to drop below 50°F, the pest control operator may heat the fumigant.

- How will your vacant home be secured? Strong warning signs with skull and crossbones will alert honest people if the tarpaulins don't, but supplemental locks should be installed on all doors. You will probably have to surrender your keys to the fumigators. Even so, you run the risk of burglary, so put your valuables in a bank vault or other safe place before the work is scheduled to begin. One reputable pest control operator had two fumigation jobs burglarized in one month. The fumigants are laced with tear gas to make them detectable, but the thieves, using roll after roll of paper towels and toilet paper to wipe their streaming eyes and noses, took what they wanted.

- How will you know if the house is safe to reoccupy? Wearing protective gear (probably gas masks), applicators open the doors and windows and turn on fans. A special device determines whether or not any gas is left. There should be none. Premature entry has caused at least one customer death.

- How long will the fumigant protect your home? As soon as the gas has dissipated, your house can be reinfested. There is no residual effect. Discuss with the pest control operator the desirability of dusting wall voids and attic with a long-lasting sorptive dust such as Dri-Die to kill any future pests thinking of setting up housekeeping. Consider contracting for regular annual inspections. Such an arrangement can be good insurance.

- How much does fumigation cost? You can probably count on spending several thousand dollars for the treatment, plus a hotel and meals for yourself and family, boarding for your pets, and a gardener to trim bushes eighteen inches away from the house.

❑ Check further. Be wary of a company that tries to pressure you. Even if you do have termites, your house is not going to tumble down tomorrow or even next week. Drywood termites develop slowly, over years. Houses have stood infested for fifty years. Besides contacting two or three customers of each company, call the state agency that oversees the industry and ask how many violations each outfit has had filed against it in the past year or two. The agency may be part of the health or agriculture department. It may be called the Structural Pest Control Board, as in California; the Office of Paints and Chemicals, as in Virginia; or something else. Be aware, though, that probably no company has a spotless record. Pest control work is complicated, difficult, and often dangerous. Human beings, no matter how intelligent and conscientious, make mistakes. Common sense says to choose the concern with the fewest marks against it. Take their relative size into account, for a small concern could have fewer violations than a large one, yet do no better a job. Further, a company may have a high percentage of error-free applications, but may be unpleasant if there is a misunderstanding about your contract. To judge the kind of people you will be dealing with, check with your community's consumer affairs bureau, chamber of commerce, or Better Business Bureau.

❏ When the applicators arrive at your home, they, their clothing, their truck, and their equipment should be clean, says Robert LaVoie, former president of a large Los Angeles pest control company. And there should be no odor of liquor about them. Termite control demands a clear head.

As with termite inspection, a professional pest control operator should provide you with a formal, written report of the findings. You should also get a completion notice of any work done.

CARPENTER ANTS

If you live in a humid area with an abundance of shade trees, you are apt to encounter carpenter ants in your home. These alarmingly large insects—up to one-half inch long—are as destructive as termites in damp regions like the Pacific Northwest. They are also abundant in southern California mountain regions.

These ants feed on dead and living insects and the juices of ripe fruit. While carpenter ants do not eat wood, they do tunnel into it to make their nests and often remove large sections of insulation and attack sound lumber. From their hideouts, they fan out at night to forage for sweets, protein, and water. Indoor nests may be satellites of a large colony outside sheltering the queen. There can be as many as twenty satellites per main colony.

You may not see the ants themselves, but if you come upon piles of sawdust near small oval slits in secluded walls, such as those found in closets, cupboards, or the attic, you should suspect carpenter ants. To locate their nests, you can set out a bait made of sugar and milk or sliced-up crickets (do you have a strong stomach?). Sometimes you can trace their paths across soil or lawns from their nest to your house. So many ants use the trail that they leave a matted road that is easily seen.

SHUT THEM OUT

As with termites, eliminating sources of moisture is a basic strategy in control, one that you should plan and carry out with a competent pest control operator. You may also need the services of a building contractor.

Who Controls the Pest Controllers?

In the United States, the federal government sets the nation's standards for pest control, and the individual states use them to certify pest control operators as competent. Pest control is a highly specialized field, requiring wide knowledge of chemistry, biology, entomology, ecology, and sound business practices. State tests cover such subjects as:

• Thorough understanding of chemical labels.

• Safety, including precautions that must be taken in treated areas; protective clothing and gear; poisoning symptoms and first aid; and proper handling, storage, and disposal of pesticides.

• Recognition and biology of pests.

• Thorough understanding of pesticides and their application.

• Proper use and maintenance of equipment.

• Proper application techniques; preventing drift and pesticide loss to the environment.

The use of state tests, however, doesn't necessarily mean that the people who come to treat your property are experts in all these areas. In some states, the applicator may not have taken any test at all, and may not even know how to read. A pest control company can legally operate with only one certified staff member who oversees those actually applying the chemicals.

Federal law requires that when a certified applicator will not be on the scene, he or she must provide detailed guidance for the actual worker. Provision must also be made for contacting the supervisor quickly.

Areas to check for moisture include clogged gutters, eaves, window frames, basements, subflooring, and porches. If moist conditions in wooden structural members of your home are not corrected, the infestation that you eliminate will be only the first in a series. Other colonies also will find your home hospitable.

WIPE THEM OUT

You may want to try to get rid of carpenter ants yourself. Look for a slow-acting ant bait with a concentration of boric acid no higher than 1 percent. Since no more than 10 percent of the colony are out foraging at any one time, it can take some time before the poison reaches the larvae and queens in the nest.

If you choose to call a pest control company, the operator will probably want to use chemical controls along with the needed house repairs. The safest controls for you and your family are dusts such as diatomaceous earth (the type used for insect control, *not* the kind used for pool filters), silica gel, pyrethrum, and technical boric acid. These are blown into the wall. Diatomaceous earth and silica aerogel, neither of which is toxic, and both of which stay in place for a long time, destroy the insects' waxy coats, causing them to lose their body fluids. Boric acid is a stomach poison. (See page 47 for a detailed discussion of boric acid.) Pyrethrins, the safest insecticides we have, may be combined with these dusts. However, if someone in your family has asthma, it would be wise to avoid this derivative of a chrysanthemum-like daisy. It can cause serious breathing problems.

Learning that your house is infested with termites or carpenter ants is upsetting. However, there's no need to panic. A competent pest control operator who is familiar with nontoxic methods of control can clear out the infestation without endangering you, while a knowledgeable building contractor can correct structural faults to prevent future damage from these voracious insects.

12. Plant Pests—
Part of the Landscape

A homeowner shows his rose garden to a friend, who admires the vigorous, fragrant blooms.

"Mine don't look anything like these," says the friend. "What systemic do you use?"

The garden's owner raises his thumb and forefinger. "These," he replies, and rubs them gently around a stem, crushing one or two aphids. "And this," he adds, pointing to a lady beetle foraging on a leaf.

His flourishing shrubs have never been sprayed with anything more lethal than a dormant oil spray or a light solution of soap. Around the entire bed, graceful onion and garlic plants stand guard. The roses get the simplest care. They are never watered after noon, and during their growing season, they receive a monthly ration of rose food.

INTEGRATED PEST MANAGEMENT IN YOUR GARDEN

This garden is a fine example of integrated pest management (IPM). Instead of insecticides, the gardener uses a number of "compatible means to obtain the best control with the least disruption of the environment," as ecologists Mary Louise Flint and Robert van den Bosch have described it. There are many ways you too can apply this new/old technology in your garden to control some of the most common plant pests.

Some IPM strategies include:

- Cultural controls, such as interplanting and crop rotation.

- Companion planting and repellents.

- Hand-picking of insects.

- Biological controls, such as natural predators and disease-causing organisms.

- Homemade plant sprays.

• Mechanical barriers.

Organic plant protection is a vast subject, and a detailed treatment of it is beyond the scope of this book. However, the suggestions in this chapter are those that experienced horticulturists have found to work against the pests most likely to be feeding in your garden. These measures should start you thinking about devising some tactics of your own to keep the hungry hordes under control without fouling the air you breathe and the water you drink.

Cultural Controls

The reason U.S. farmers rely heavily on chemical pesticides is that they plant broad spreads of a single crop, which builds up a great potential for that plant's pests. As a home gardener, you don't have to do that. Vary your crops as much as possible.

Interplanting

By setting out your plants somewhat at random instead of in tidy rows, you will make it harder for pests to find a continuous food supply. Chinese farmers employ this method, planting some wheat, with some potatoes nearby, and maybe some rice, so that insects that feed on any one of these crops are confined to a small area.

Crop Rotation

If you plant tomatoes in one spot for several years, tomato pests have a chance to build up their populations. You also deplete the soil of nutrients, so that you have more pests attacking weaker plants. To avoid this, plant a different crop in the same location each year, such as tomatoes one year, squash the next, and peas the following year. Or consider letting your flower and vegetable plots trade places.

In choosing vegetables, try to avoid pest-prone plants like cucumbers, melons, radishes, and broccoli. Instead, use such crops as beans, peas, and spinach—all hardier species.

It is a good idea to put in more vegetables than you really need so that you can tolerate some loss to insects or plant disease. You can also time your plantings to avoid a particular crop's pests. Your county farm or cooperative extension agent can advise you on this, as well as on proper watering and fertilizing practices.

Once you have brought in your harvest, pull up all the plants that are past their growing season so that their pests will have no winter harborage and so will die. In the fall, turn the soil over to let the birds deal with grubs and maggots. Those that the birds miss will die of starvation or cold, or will simply dry up.

Consider letting a small stand of weeds in an out-of-the-way corner stay for the winter. It will serve as harborage for a few pests, enough to keep the predators around for the next season.

Companion Planting and Repellents

Cultural controls are like housekeeping transferred out of doors. The same strategies of rotation, plenty of fresh air, and cleanliness work as well outside as they do inside. These methods, says the National Academy of Sciences, "are often economical and dependable, although seldom spectacular." In some parts of the world, farmers use them as a matter of course. They also use a strategy called companion planting.

In Russia, farm workers set mustard plants among their cabbages to increase the effectiveness of parasites that live on cabbage worms. In Hawaii, growers surround squash and melon fields with a few rows of corn, which is very attractive to the melon fly, a major pest. The corn is a trap crop that attracts pests away from the crop you are trying to protect.

Onions and garlic growing here and there throughout your garden will repel insects. Allyl, an active component in these plants, has been found by University of California scientists to be an effective mosquito repellent. Strong-smelling herbs like rosemary, peppermint, pennyroyal, southernwood, wormwood, lavender, and sage planted among crops are also said to hold off insect pests.

Hand-Picking of Insects

Although too time-consuming to be practical on a commercial farm, the hand-picking of insects like hornworms on tomatoes or bagworms on ornamentals is feasible for the home gardener. You can either gather the pests in a plastic bag and dispose of them with the garbage, or put them in a jar with a little water, where they will decompose. When the same species appears again, set the open jar under the plant they're invading. This can be a strong repellent, because most animals tend to avoid their own waste.

Biological Controls

There are two primary approaches to the biological control of pests. The first is through encouraging the pests' natural predators, which feed on the undesirables. The second involves using natural biological organisms that kill pests by causing disease in them.

Predators

In holding down the garden pest population, you have some formidable allies whose services cost you nothing and who never pollute the environment. They don't have to be sprayed or dusted on, and they often go to work long before you're even aware that you have a pest problem. These are the predators—birds and insects, lizards and toads—that feed on the pests that feed on your plants.

Insect predators are the world's most thorough insecticides. Many pest young—eggs, larvae, and pupae—have become immune to chemicals, but a predator can swiftly lay its egg on the young of another insect so that the newly hatched predator consumes its hapless victim long before the latter has a chance to do any damage. And the best part is that these predators cost you nothing. Among the most helpful insect predators—all found in urban areas, as well as in the countryside—are lady beetles, dragonflies, lacewings, hover flies, ground beetles, centipedes, and parasitic wasps, which include the miniature-sized Trichogramma wasps, the larger ichneumons (also known as ichneumon flies), and others. Larger predators include snakes, toads, lizards, hawks, owls, ravens, seagulls, weasels, bats, racoons, badgers, and opossums. Some would add praying mantises to this list. However, praying mantises are not really all that beneficial since they eat everything in sight, including each other and truly helpful insects.

Figure 12.1 shows a few of the most helpful garden insects.

Natural predators like lady beetles, Trichogramma wasps, and lacewings can be bought from commercial insectaries for release on your property. Reviews on

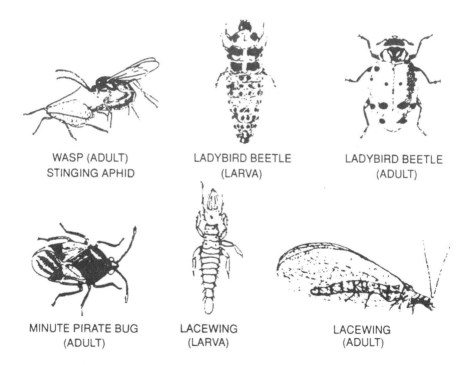

| WASP (ADULT) STINGING APHID | LADYBIRD BEETLE (LARVA) | LADYBIRD BEETLE (ADULT) |

| MINUTE PIRATE BUG (ADULT) | LACEWING (LARVA) | LACEWING (ADULT) |

Figure 12.1 Beneficial Garden Insects

Source: Controlling Insects, Diseases, and Related Problems in the Home Vegetable Garden (Division of Agriculture Sciences, University of California, Leaflet 21086, 1979).

this have been mixed. The immigrants don't always take to their new surroundings and may not give you the hoped-for results. (See the inset on page 212 for tips on making your garden a more hospitable environment.) As with all aspects of pest control, nothing is guaranteed. However, mass releases of commercially bred predators have been successful in Great Britain when aimed at aphids and spider mites in greenhouses; and have also been effective in the United States against citrus pests. In the late nineteenth century in California, a scale insect was destroying the state's young citrus groves. The vedalia beetle, a kind of lady beetle, was imported from Australia, where it preyed on scale insects. The beetle flourished in California, and within two years had almost completely routed the scale insects. Then, in the late 1940s, DDT was introduced as the ultimate insecticide, and California citrus growers adopted it as enthusiastically as other farmers did. Along with various pest species, the beneficial vedalia beetle was also eradicated—and the scale insect rebounded with devastating effect. The damage to the groves was the worst it had been in fifty years. We have now come full circle, with many growers not only trying to protect natural predators but often releasing batches of parasites and predators purchased from commercial insectaries.

The story above demonstrates a major drawback of the use of insecticides: They often kill beneficial insects as well as harmful ones. Thus, though they may bring immediate relief, in the long run there is a serious risk of an upsurge in the population of a secondary pest that was of no importance before but that, once free of its enemies, multiplies astronomically.

One problem for the home gardener practicing biological control is that while you are wooing the predators, pesticides may be drifting over from neighbors' yards. If you can enlist the community's cooperation to hold off on the chemicals and give the predators a chance to do their work, you should all enjoy a more healthful environment.

Disease-Carrying Pest Controllers

Organisms that cause insect diseases are also potent pest controllers. The most common, *Bacillus thuringiensis*, or BT (sometimes called BTI), is available in several varieties as an insecticide in garden shops. It is no threat to mammals, and usually poses no problems to helpful insects.

Another disease agent, milky spore, was a major factor in bringing the Japanese beetle to heel in the 1940s and 1950s. It is still used, and is available in garden centers.

Homemade Plant Sprays

Organic gardeners have long known that homemade sprays can be of great help in protecting plants. The simplest is clear water. Often, a strong spraying from a garden hose is enough to dislodge plant pests.

Attracting Beneficial Insects to Your Garden

As explained on pages 209 to 211, a gardener has many natural allies eager to wipe out the destroyers—but gardens must be made welcoming to these help-ful predators. Beneficial insects require enormous amounts of energy to lay eggs and find pests to eat. When these "good" insects are moved to a new area, flowers are often the first things for which they look. Pollen (protein) and nectar (carbohydrates) provide the energy they need and serve as an alterna-tive food when the insects on which they feed are scarce. Flowers with hori-zontal clusters, like wild buckwheat, provide accessible landing platforms. Annuals like alyssum and white lace also attract these beneficial insects.

Don't expect results the moment your plants begin to flower. Most of the time, you will be feeding the parents of the beneficials. Their offspring will slowly build in number, and you will eventually begin to see results. The following table directs you to those plants that will attract the six-legged friends in which you're interested.

Plant These ...	To Attract These ...	To Feed on These ...
Caraway, cilantro, dill, fennel, sweet alyssum, tansy, and white yarrow.	Green lacewings, parasitic wasps, syrphid flies, and tachinid flies.	Aphids, caterpillars, mealybugs, mites, thrips, and whiteflies.
Flowering buckwheat (*Fagopyrum* sp.) and wild buckwheat (*Eriogonum* sp.).	Green lacewings, ladybug beetles, parasitic wasps, syrphid flies, and tachinid flies.	Ants, aphids, gnats, houseflies, maggots, moths, slugs, snails, and termites.
Evening primrose.	Ground beetles.	Aphids, caterpillars, mealybugs, mites, thrips, and whiteflies.
Clover (crimson, white).	Ladybug beetles and parasitic wasps.	Aphids and caterpillars.
California lilac (*Ceanothus* sp.) and Baby Blue Eyes.	Syrphid flies.	Aphids.
Coyote bush (*Baccharis* sp.).	Green lacewings, parasitic wasps, syrphid flies, and tachinid flies.	Aphids, caterpillars, mealybugs, mites, thrips, and whiteflies.
White lace flower and yarrow.	Damsel bug, hover flies, ladybug beetles, and parasitic wasps.	Aphids, caterpillars, leaf hoppers, and thrips.

Source: "What to Do When the Bugs Come for Your Garden" (Community Environmental Council, Santa Barbara, California, 1995).

Soap sprays have been used successfully since early in the nineteenth century, when fish- and whale-oil soaps were the rule. Modern biodegradable soaps or detergents, at a concentration of 1 or 2 percent (roughly one or two tablespoons to one gallon of water), also work.

In tests conducted by the California Department of Agriculture, liquid soaps and detergents provided good control, in some cases comparable to that of synthetic organic pesticides. Mites, aphids, psyllids, and thrips were all killed by the sprays, which seem to work in two ways: by dislodging the pests, and by smothering them. Ivory liquid detergent or Ivory flakes gave the most consistent control. There are also several commercially available soap sprays designed for the home garden or house plants. The scientists suggest that you rinse the plants off with clear water several hours after the soap spraying to reduce leaf burning.

Mechanical Barriers

A number of mechanical barriers can keep crawling pests out of shade and fruit trees. All of the following have proven effective:

• The National Academy of Sciences (NAS) recommends wrapping sticky bands like Tanglefoot around the trunk of the tree, two to four feet from the ground to block cankerworms, gypsy moths, cicadas, and ants who pasture their aphids in trees.

• The U.S. Department of Agriculture says that applying flypaper to a band of heavy paper six to eight feet wide and then tacking the band around the trunk also works. Fill the rough places in the bark with cotton batting so that pests can't crawl under the paper.

• Another barrier is cotton batting several inches wide wrapped tightly around the trunk. Tie the batting tightly with string near the bottom edge and a couple of inches from the top of the band. Turn the upper portion down over the lower to trap crawling pests.

• Mosquito netting, sixteen wires to the inch and fourteen inches wide, can also guard a tree. Tack it to the trunk so that it fits tightly at the top and flares out half an inch or more at the bottom. Smooth or fill all rough places on the bark so that caterpillars can't crawl under the top edge of the band.

CONTROLS FOR SPECIFIC PESTS

The following nonchemical controls have been found to work for some of the more common garden pests. You may have to try several before you achieve the results you want.

Aphids

Most plants can sustain a moderate number of aphids without harm, so finding a few on your favorite roses is no reason to bring up the chemical howitzers. Steps you can take to minimize these tiny, pear-shaped pests include application of sticky bands or petroleum jelly, soap sprays, rubbing, and pruning.

Ants fight off aphid predators to keep getting honeydew from aphids, scales, and mealybugs, so keeping ants away from their "cattle" gives the predators a free hand. Sticky bands wrapped around plant stems will do just that. When using petroleum jelly to protect your roses, use a sturdy brush to apply the goop—there's no need to risk being raked by thorns. Aphids are food for a number of other insects and their young, including the well-known lady beetle, lacewings, syrphid flies, daddy longlegs spiders, and praying mantises. Birds also find them tasty. Although aphids' reproductive rate is awesome, predators and natural forces like cold rains, temperature extremes, fungus, and bacterial disease hold their numbers down.

You can help keep the aphid population within tolerable limits. Soapy water, as mentioned, sprayed at a firm pressure washes aphids off plants or smothers them. Water alone can also do it, but a little soap increases its effectiveness. Or you can rub infested plant parts gently between your thumb and forefinger, or wipe them with a damp cloth. Pruning infested plant parts also helps. On sensitive new growth, pick off the pests with a moistened cotton swab or a toothpick.

To control aphids on trees, thin out the dense inner canopy that provides the pest with protected living quarters. When forced to the tree's outer growth, aphids are much more susceptible to cold and wind, as well as to predators.

If you are thinking of buying and releasing a gallon or so of lady beetles, the California Cooperative Extension warns that this seldom gives successful control. The insects, which are collected from their hibernating places in the mountains, fly all over the area. "None," says entomologist Walter Ebeling, "will stay in the garden."

Cutworms

These nuisances, the larvae of certain moths, destroy young plants by nibbling at a seedling's base. You can take several tacks to get rid of them. An inverted cabbage leaf set near infested plants will trap the cutworms at night. Hand-picking at night, with the aid of a flashlight, is another tactic. If that's inconvenient, ring your plants with empty pet food or tuna cans with both ends sheered off. Sharp-edged materials like diatomaceous earth or crushed eggshells can also protect your young plants from cutworms. Pushing a small twig or nail into the ground right next to a seedling will supposedly keep a cutworm from wrapping itself around the young plant and killing it. It could be worth a try.

Because cutworms are addicted to cornmeal but cannot digest it, it is lethal to them. Some gardeners recommend sprinkling the grain around seedlings; however, rats and mice also like cornmeal, so you may be solving one problem and setting yourself up for a worse one if you do this.

Earwigs

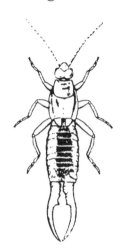

Figure 12.2
European Earwig
Source: Earwigs and their Control (Division of Agricultural Sciences, University of California, Leaflet 21010, 1982).

Earwigs are not necessarily pests. Actually, most are beneficial. If you find one snuggled among some rose petals, it just might be eating aphids. Sometimes they do eat plant material or ripe fruit, though. To make sure that they are the culprits in your garden, check your damaged plants at night with a flashlight to see if an earwig is doing the damage. (See Figure 12.2.) If they get into a house, their foul odor makes them an unpleasant nuisance.

One of the best controls for earwigs is trapping. Use rolled-up newspaper, a section of old garden hose, a length of bamboo, or a rolled-up burlap bag. Set the traps around the garden where you have seen the insects, and pick them up every day or two. Then knock the earwigs into a pail of soapy water, or just crush them.

Can an insect be endearing even if you're not an entomologist? You decide: The female earwig has a marked mothering instinct, protecting her eggs and the early molts of her young from predators. But you certainly don't want them around *your* young, for they can give a painful pinch. If you think that some might be hiding in your children's sandbox, sift the sand through window screening to trap them. Crush any you find.

Gypsy Moths

Gypsy moths are a major forest pest in the eastern United States where, during their peak population years, they defoliate millions of trees. Few of the trees die, however. Decades of intensive pesticide application have not curbed this insect, which now pervades millions of forest acres. In the past, the pesticide carbaryl (perhaps better known as Sevin) was sprayed against the moths, but by killing off beneficial insects, the chemical may have compounded the problem. By the late 1980s, nontoxic BT was the agent of choice in most moth-infested states. Dimilan, a growth regulator, was also being used.

You can protect your garden with banding, as described under Mechanical Barriers on page 213. You can also apply BT to infested trees. Watch for the moth's egg masses and destroy them. They are generally found on outdoor items like garden furniture, toys, camping equipment, and building material. Incidentally, if you are planning a move to a moth-free state from an infested one, federal regulations require you to have all outside household articles inspected before the move and certified as free from gypsy-moth caterpillars, eggs, and cocoons. Quarantined states include Connecticut, Delaware, Illinois, Indiana, Maine, Maryland, Massachusetts, Michigan, New Hampshire, New Jersey, New York, Oregon, Pennsylvania, Rhode Island, Vermont, Virginia, West Virginia, Wisconsin, and the District of Columbia.

If you are moving from a quarantined location to one not infested with gypsy moths, you can have your articles certified in one of three ways:

1. The shipping company may give you a self-inspection kit that includes an appropriate document for you to complete and give to the van driver.

2. You may hire a private pest control company to check your property and certify it.

3. Your state agriculture department may certify that you are not carrying any gypsy moths with you.

Japanese Beetles

The Japanese beetle is a beautiful insect with bronze wing covers and a metallic green body rimmed with small patches of white hair. It was accidentally introduced into the United States in 1916. Without natural predators, it quickly fanned out from its arrival point, and is now found in just about every state east of the Mississippi. Although still a garden nuisance, it is not nearly as destructive as it once was, thanks to an extensive federal program of spraying milky spore bacteria, which is fatal to this beetle. The bacteria persist after application and ultimately spread to other areas.

A daytime feeder, the Japanese beetle is most active on clear, warm days and in bright, sunny areas. These habits make hand-picking easy, since the insect is sluggish in the morning's cool air.

To trap Japanese beetles, spread a large cloth under infested bushes before 7:00 a.m. and shake the insects from the plant. Then gather them up and drown them in warm, soapy water. You might also try a commercially available trap called Bag-a-Bug, which uses sex pheromones to lure its victims.

Natural predators of Japanese beetles include chickens, ducks, turkeys, pigs, grackles, starlings, meadowlarks, cardinals, and catbirds. Shrews and skunks also fancy them.

Snails

This universal botheration was introduced deliberately into the United States by French immigrants as a delicacy. Our palates betrayed us, for snails are among the worst ravagers of field and garden. A few species prey on small insects, and one, the decollate snail, is a cannibal, feeding on its fellow mollusks. Nearly all, though, are extremely destructive.

Snails are most active at night or in damp, foggy weather. When the thermometer falls below 50°F, they are inert. On sunny, dry days they hide under boards and rock piles and in low-lying shrubs and dense vegetation. They tend to favor beds of ivy. Long dry spells don't kill them; they just seal themselves up in their shells, and can remain dormant for as long as twelve years.

Metaldehyde, the active ingredient in most snail baits, is a strong poison, but snails are becoming resistant to it. To make matters worse, the bait is often extremely attractive to dogs, who will tear a box open to get at it. Metaldehyde can throw a dog into convulsions or even kill it. Nor should you keep metaldehyde around if you have small children. The pink granules don't look all that different from some candied dry cereals.

Actually, there are so many ways to control snails—garden sanitation, traps, barriers, hand-picking—that there is really no need to use poison baits.

Garden Sanitation

To prevent or get rid of snails nonchemically, do the following:

❑ Clear your yard of unused boards, stones, and other debris.

❑ Repair any leaky outside faucets or hoses.

❑ Clear out weeds and other unnecessary foliage so that the soil dries more quickly and makes a less hospitable snail environment.

❑ Remove dense ground covers like ivy. (If you must keep the ivy, set your snail-susceptible plants as far as possible from it.)

❑ Protect low-growing plants (snails favor them) by ringing them with an inch or two of wood ash, sawdust, lime, sand, or diatomaceous earth. All are very irritating to mollusks. Since these barriers are less effective when wet, you must replace them after a rain. Spread the irritant around the perimeter of vegetable plots to keep snails out of the whole area. You may have to check with a lumberyard to find sawdust. Some of them may give it to you free if they have it.

❑ Hand-picking is probably the fastest way to clear out snails. Let the soil get very dry and then, late in the afternoon, turn on the sprinkler. After dark, go out with a flashlight to gather the snails (there will probably be dozens) awakened by the moisture. (If you'd rather not touch them, you can use kitchen tongs to pick them up.) Put them in a bag and crush them. Since they are rich in nitrogen and calcium, bury them for fertilizer.

❑ If the weather is so damp that the soil doesn't dry out, hand-pick whatever snails you find in the morning. At first, hand-pick every day; but as their numbers dwindle, a weekly gleaning should be enough.

❑ Sprinkle salt heavily on a snail that is out of its shell to kill the animal by drawing out all of its body fluids. Be aware, however, that salt can damage soil, and so should be used only on pavement. Stepping on the pests also seems simpler if they are on a walkway.

❑ Make use of snail predators, which are many and diverse, from lightning bug larvae to human beings. Among the most voracious are ducks and geese, but garter and grass snakes, box turtles, salamanders, some toads, and an occasional dog (possibly with French poodle ancestry?) will also eat them. Rats also feed on them, which is a good reason to keep your garden's snail population down.

❑ If you want to become a snail predator yourself, feed those you have gathered with cornmeal or oatmeal for at least three days to clean and purge them, find a good recipe, and invite your friends for an escargot party!

Traps

Employ any number of simple traps that are useful for controlling snails in an otherwise clean garden.

❑ An easily made board trap, set on runners, attracts snails by giving them a shady cover. (See Figure 12.3.) Just be sure to empty it every morning, or the snails will crawl off and continue their gorging.

❑ Another effective trap is a saucer of stale beer mixed with a bit of wheat flour to make a stickier brew. Very thin fermented bread dough will also catch them. Raw slices of turnips or potatoes will bring them out, and flowerpots and grapefruit shells turned rim to the ground are said to provide ideal snail hideaways and, hence, traps.

❑ A third trap, this one suggested by *Sunset Magazine,* consists of two empty slightly tapered one-gallon nursery cans with drainage holes. Nest one inside the other, leaving a gap between the two bottom panels. Set the trap on its side in a spot that is shaded all day long. The slugs and snails will crawl into it through the outer can's drainage hole. Be sure to empty the trap every day.

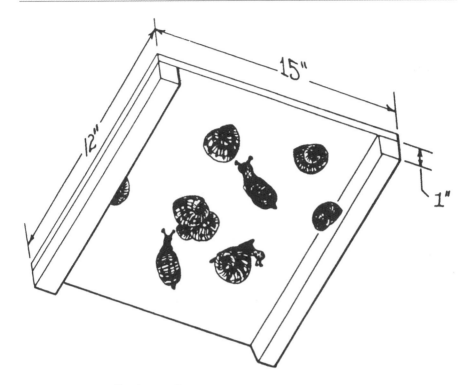

Figure 12.3 An Effective Snail Trap

Boards on 1-inch risers can be placed in the garden to trap snails. Snails should be removed and disposed of each morning.

Source: Snails and Slugs in the Home Garden (Division of Agricultural Sciences, University of California, Leaflet 2530, 1979).

Sowbugs and Pillbugs

Sometimes called wood lice, these tiny creatures are not really bugs at all. They are related to lobsters, crabs, and other crustacea. For the average person, the only perceptible difference between the two is that the pillbug curls itself up when threatened and the sowbug does not. (See Figure 12.4.)

Overall, sowbugs and pillbugs are useful scavengers that turn decaying organic matter into soil nutrients. Occasionally, though, they may injure a young plant. If there are too many of these creatures in your yard and they are getting into the house, try to eliminate the damp conditions they like by doing the following:

❑ Remove all boards and boxes in the garden.

❑ Eliminate piles of grass, leaves, and other hiding places, especially those next to the foundation of the house.

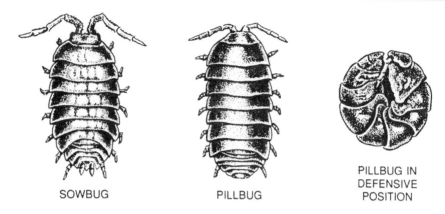

Figure 12.4 A Sowbug (left), a Pillbug (center), and a Pillbug in the Typical
Defensive Position (right)

Source: Sowbugs and Pillbugs (Division of Agricultural Sciences, University of California,
Leaflet 21015, 1980).

❏ Plant shrubs and flowers far enough from the house so that wind and sun can
keep the foundation area dry.

❏ Keep foundation and sills tightly caulked.

❏ Ventilate crawl space and basement adequately.

❏ Don't overwater the garden.

Spider Mites

"When you kill off the natural enemies of the pests," said the late scientist Carl
Huffaker, "you inherit their work." Spider mites are a prime example of this. Once
minor pests, since the introduction of synthetic organic pesticides, which wipe out
mite predators, they have ballooned into a serious problem worldwide. Because
the pest's predators—lady beetles, lacewings, flower bugs, and predaceous
mites—take longer to reproduce than it does, killing them off leaves the spider
mite free to reproduce at an enormous rate.

Signs of spider mite infestation are leaf damage, webbing, bronzed or stippled
leaves and stems, poor leaf and flower color, fewer fruit, and buds that fail to open.

Applying a dormant oil spray, available at nurseries and garden centers,
blocks the mite's population buildup. The U.S. Department of Agriculture recommends hosing infested plants with a strong spray of soapy water. In warm
weather, the spray should be applied several times a week. Be sure to reach all
parts of the plant, including the undersides of leaves.

Whiteflies

These tiny creatures are most often seen in the garden in their adult form. They particularly like to rest and feed on the undersides of leaves. When an infested plant is disturbed, the mothlike creatures will fly off in clouds. The yellow sticky board traps sold in nurseries can be effective in controlling a greenhouse infestation, but in open garden spaces they do little good. Cold water from the hose, directed at the undersides of leaves, kills whitefly larvae hovering there.

PEST CONTROL FOR HOUSE PLANTS

House plants may be more susceptible to insect pests than shrubs or flowers. To treat a plant that is heavily infested with any of these pests, cut away the blighted parts and dispose of them with the garbage. Isolate the plant for a few weeks.

Some house plant problems may not be caused by insect pests at all. Inadequate light, poor air circulation, sudden bursts of cold air, and water-sprayed leaves can all damage tender tropicals. So check these factors before treating the plants for what may be nonexistent insects.

If you do find a few mealy bugs or scale insects on your house plants, wet them with an alcohol-dipped cotton swab for a quick kill.

If we can accept insects and other small wildlife as integral to our home's landscape, we will be more willing to share space with them and more reluctant to call out the troops at the first sign of an insect, beneficial or otherwise. Healthy plants can tolerate considerable damage without keeling over, so before turning to chemical pesticides, try some patience, a bit of ingenuity, regular maintenance, and a willingness to accept a less than picture-perfect garden.

PART FIVE

Controlling the Controllers

13. Ants, Spiders, Wasps, and Scorpions— Useful but Unwelcome

The ant, the spider, the wasp, and the scorpion—each is both a help and a pest. When they burrow in the soil, enriching and aerating it, or when they attack termites and other insects that harass us, ants are our welcome allies. But let us find them swarming over the kitchen countertop after a hard rain and we want them out—yesterday. Seeing a spider's web stocked with dead houseflies and gnats, we are grateful that these disease-carriers have been dealt with for us. Nevertheless, even a small web in the corner of a room's corner sends a careful homemaker flying for a broom. We may be only vaguely aware of what wasps and scorpions do that benefits us, but we say, "Please! Let them do it somewhere else!"

Scientists who spend their entire lives studying these creatures marvel at their abilities. A spider's web is an engineering miracle, exquisitely suited to providing its builder with food. Some species of wasps make their superbly crafted nests of paper so sturdy that you can write or even type on it. And ants set up their social systems millions of years before humans appeared.

However astonishing their talents and however great their benefits, when ants, spiders, wasps, or scorpions get too close for our peace of mind, most of us want to get rid of them firmly, safely, and as quickly as possible. Let's consider ants first, since they are the creatures most likely to move in with us and bring all their coworkers.

ANTS

Ants are found almost everywhere on earth, from snowy mountaintops to steaming swamplands. They surpass in numbers all other land animals. Their great underground cities, said naturalist Edwin Way Teale, can outlast a human generation.

Ant societies are organized by castes, each of which has well-defined tasks. Queens and males are responsible for the group's continuity; and female and neuter workers, for its day-to-day survival. Workers forage for feed, keep the eggs and cocoons clean, feed the larvae, and defend the nest. The same keen sense of smell that brings them to the jam you forgot to wipe up yesterday tells them who is a member of the colony and who isn't. In many species, an ant that doesn't smell like the rest of the colony is barred from the nest.

If a potential predator shows up, the workers squirt it with formic acid—a form of chemical warfare at which they are expert. They can shoot the acid up to a foot away, a huge distance for a creature that is only one-tenth of an inch long.

Ants resemble termites in some of their habits. Both species are social, live in large colonies, and mate in huge flying swarms. Both drop their wings and nest in hidden places—underground, or deep within wood or a decaying tree. Both can form enormous communities, with up to a quarter-million individuals in one nest. Egg-laying ant and termite queens are imprisoned for life in their chambers and must be cared for by the workers. In contrast to those of the more secretive termites, ant colonies are often quite visible, with workers hurrying back and forth on important errands.

What Ants Eat

Whereas termites eat only wood-related products, ants will eat just about anything humans do. They will devour sweets, fats, starches, grains, and meats. When seeking meat, ants, in large numbers, can kill a young bird or the hapless young of just about any species. They can swarm and devour lizards, geckos, and other small animals actually beneficial for the environment.

The Benefits of Ants

Ants generally do more good than harm. The nuisance that ants can be is relatively slight when weighed against their many benefits to us. Besides enriching the soil, ants prey on cockroaches, conenose bugs, the larvae of filth flies, and other pests. They will even attack the Oriental rat flea, bearer of plague.

In thirteenth-century China, orchardists used warlike ants to guard their orange trees from citrus pests. Today, German law protects red ants so that they, in turn, can defend the nation's forests. In cotton-growing regions of the United States, ants help control the boll weevil. One of ants' greatest benefits to us is their fondness for mosquito eggs, some species of which can stay in a dry spot for years until a rain or other watering sets them to hatching. Foraging ants often prevent this from taking place.

Ants can be good termite insurance, for the two are age-old enemies. Because ants will attack a wingless queen termite searching for a nest site, they probably prevent much termite damage to our homes.

Their benefits aside, however, an infestation of ants in the home is far from pleasant.

How Ants Make Themselves at Home

In earning their living, ants show remarkable adaptability. The harvester ant nips the seeds that it gathers into its nest to keep them from sprouting. After a rain, workers carry the seeds to the surface to dry them out and prevent them from molding. Another species, the leafcutter, carries bits of leaves underground to form a fungus garden, which it fertilizes with its own excrement. Ants that tend aphids sometimes build paper shelters over their "cattle" to hide them from predators. They have also developed war to a high art. Winners of a battle carry home the cocoons of the conquered. The hatchlings become slaves, doing all the work formerly done by the victors, who ultimately may pay a high price for their victory by forgetting how to take care of themselves.

Ants are most apt to invade a home after their nest has been flooded. It does not matter if the deluge is caused by heavy rain or just garden watering. When our son repotted some houseplants outside the house and forgot to turn off the garden hose, an hour later the outside wall was covered with thousands of tiny confused, suddenly homeless creatures. Their scouts soon found a crack or two in the wall and signaled to the whole tribe that delicious tidbits awaited on the dirty dishes in our dishwasher. They also found a dab of dried jam in one of the cupboards. It took us a week to persuade them to find another place to live. As we discovered, ousting ants takes ingenuity and patience—and a sense of humor. We did finally succeed, and without chemicals.

SHUT THEM OUT

It is much easier to keep ants out than to try to evict them once they have found their way in. Here are some simple preventives:

❏ Trim all tree and shrub branches well away from the house, since ants can crawl along them and get in through window and door cracks. Rake mulch away from the house.

❏ Patch wall cracks both inside and out.

❏ Don't overwater plants or your garden, especially near the house.

❏ Wipe up spilled foods immediately.

❏ Rinse dishes well before putting them into the dishwasher; or wash them by hand after each meal.

❑ Store all foods in tightly closed containers.

❑ Wipe clean the outside of all bottles and jars of jam or syrup, including liquid medicine, immediately after using them.

❑ Wrap leftovers, especially cake and cookies, closely in foil or plastic, or put them in sealed plastic containers.

❑ Sweep, mop, or vaccum your kitchen floor daily to pick up those tiny fallen scraps.

❑ If you drink a lot of coffee, try a ploy the French use: Spread damp coffee grounds around the perimeter of your house to discourage ants.

GET THEM OUT

If ants have already made their way into your home, use the following measures to evict them:

❑ Trace them back to their entry point and seal the hole with petrolatum (petroleum jelly), putty, or plaster, says William Olkowski of the Center for the Integration of Applied Science in Berkeley. Petrolatum gives very quick results, but for a more permanent effect, use the longer-lasting materials. You probably will not get complete control right away, but persevere. Eventually, you will close off so many of the ants' entries that the few stray crumbs around your kitchen will be too far from their nests to be worth their while.

❑ Foraging ants are the colony's senior citizens. They are considered expendable because their lives will soon end. To discourage any followers, wash their scent trails with a good household detergent to destroy the chemical odors guiding them. Sometimes this keeps the horde at bay . . . sometimes it doesn't.

❑ Vacuum or wipe up any stragglers still wandering around indoors. These may be scouts who will bring others if they find food. Be sure to empty the vacuum cleaner bag outside, or ants may come crawling out of it to pester you all over again.

❑ These energetic foragers can get in through surprising crannies—a crack around a window frame, a circuit breaker with tiny crevices around the wires. We once found them coming through a light switch. A dab of petroleum jelly embalmed the one poking its head out alongside the button and kept out its mates.

❑ The quickest way to deal with an ant swarm in your dishwasher is to run the machine.

❑ Mint and other herbals are often recommended as ant repellents. I have put freshly picked mint and tansy in a kitchen cupboard during an ant invasion. The insects stayed away from that cupboard for about an hour. However, not until every sweet spot was cleaned up and every wall crack was closed did I finally win control. It's the house cleaning that goes along with spreading the "repeller," according to expert Robert Wagner of the University of California, Riverside, that really does the job.

WIPE THEM OUT

❏ If your efforts in the house are not giving you enough control, you can destroy the nest outside. If you can actually find the nest, pour boiling water or hot melted paraffin (be careful!) down the nest entrance. Since ants can survive for days under water, flooding the nest with cold water will not do the trick. (If you are a supporter of an organization such as People for the Ethical Treatment of Animals, though, your conscience may not allow you to do this.)

❏ Because many ant species are shallow nesters and most dehydrate quickly, you can eliminate nests you discover under places like stepping stones, rocks, and logs by merely turning them over so the ants are exposed to sun and drying air.

❏ If you can see where the ants are entering, spraying a few blasts from a liquid window cleaner (which contains a harmless repellent) or from a mild cold-water detergent directly on the insects will discourage them. Both will kill the foragers while destroying their scent trail, which is what draws their followers. You may have to do this several times, but this safe and effective method often works. (It's always my first tactic.) Simple Green and Windex are examples of products that are suited to this purpose. You may get a complaint or two from a family member with an unusually sensitive nose, but you are not putting anyone in harm's way with this method.

❏ Employ an effective ant bait.

   • I have found the following bait—adapted from one recommended by biologist Helga Olkowski of the Center for the Integration of Applied Science in Berkeley, California—to be effective against ants infesting a house. Unlike some other baits, this one seems to be fairly quick acting. Here is the formula:

   1. Mix 3 cups of water with 1 cup of sugar and 4 teaspoons of technical boric acid that is formulated for pest control. (For a full discussion of technical boric acid, see pages 47 through 50.)

   2. Wrap 3 or 4 jam jars with masking tape.

   3. For a fairly widespread infestation, loosely pack each of the jars with the type of absorbent cotton sold for first aid purposes, and pour $1/2$ cup or so of bait into each jar.

If you have young children or pets, screw the lids tightly onto the jars and seal with adhesive tape. Pierce the lids, making two or three small holes, and smear the outside of the jars with some of the baited syrup.

If the baited jars will be out of the way and if you have no children or pets, leave the jars open. Place the jars where the ants are foraging. If you do have young children or pets, be sure to place them where only the ants will be able to get to them. It may take a few hours or a few days, but the insects will eventually swarm to the jars. Some will die near the bait, but most will carry the sweet poison back to the nest, where it will wipe out the colony. Don't kill the massed insects; they are efficient distributors of this insecticide. And, if you have patience, ant workers will return to consume their dead.

A woman I know who for years had had a stubborn bathroom infestation of ants—one that no pest control operator applying conventional pesticides had been able to dislodge—used this bait with great success. "It's a good thing I had a second bathroom," she told me, "because my bathtub was black with ants for a week." Since using this bait, she has had few if any ant problems.

• Texas Cooperative Extension, a part of the Texas A & M University system, recommends using any of the following homemade baits:

1. 1 tablespoon of peanut butter, 1 teaspoon of brown sugar, $\frac{1}{2}$ teaspoon of boric acid.

2. One 8-ounce jar of mint apple jelly, 2 tablespoons of boric acid.

3. Two teaspoons of boric acid dissolved in $\frac{1}{4}$ cup of hot water, 1 cup of light or dark corn syrup.

To assemble the bait, mix all the ingredients together. It should be a thick, sticky texture (if it is too sticky to spread, gradually add a bit of boiling water until the proper texture is achieved). A good method of putting out the bait is on masking tape. Stick a small piece of the masking tape in places where you have seen ants and spread some of the bait on the tape. Change the tape every two or three days. Ants switch food every two or three days depending on the number of developing young in the nest, so it is a good idea to alternate between a sugar-based bait and a protein-based one. It may take up to six weeks for control, but results are sometimes seen in seven days. Again, only use technical boric acid for trapping.

Using Insecticides

Baits containing arsenic, once widely sold for ant control, can still be found on garden shop shelves, as well as in homes. Don't use them. Arsenic readily combines with common elements like hydrogen and oxygen to form extremely poisonous substances. It also remains indefinitely in the ground, never decomposing. In addition, some of the baits are not in childproof containers.

Other chemicals can also cause harm. When a young woman developed a grave form of anemia, her doctors suspected the cause to be a household pesticide she had sprayed heavily in her small kitchen to kill a stubborn ant infestation. If your problem is so bad that you are considering using poisons, you would be much better off to hire a competent pest control operator to clear up the situation quickly. Incidentally, if you spray an insecticide on an infestation, you could cause the colony to "bud"—members of the original colony will scatter to form new colonies—and eventually there will be more colonies to harass you.

Future Controls

University of California scientists have found that soldiers of several species of desert termites produce a powerful ant repellent. The biologists hope to reproduce the repellent synthetically, and foresee millions of United States homes ringed by the substance while the ants march peaceably away. Since ants kill termites, however, keeping them away from a house could prove to be a double-edged sword. Which would you rather take a chance on—ants or termites?

SPIDERS

On a sunny day in 1969, a group of citizens gathered near the entrance to the city park in the small town of Sierra Madre, about twenty miles northeast of Los Angeles. Television crews bustled about, focusing their cameras on a brown metal object atop a pedestal.

The center of the fuss was a statue of a spider. It may have been the world's only monument to a spider, but this species, the "violin" spider—one of the most dangerous—had put that small community in the spotlight, so it well deserved the honor.

The statue marked the first time an infestation of the violin spider, also known as the South American brown spider and the brown recluse spider, had been seen on the West Coast of the United States.

Spiders have fascinated people since time immemorial. The Greeks told of Arachne, so proud of her beautiful weaving that she challenged the goddess Athena to match her fabrics. Offended, the goddess destroyed the mortal's work.

In despair, Arachne hanged herself, whereupon the compassionate Athena changed her into a spider.

The Navajo say that the spiders taught them to weave, and, to this day, their grateful weavers leave a tiny hole in their blankets to recall the entrance to a spider burrow.

Scottish legend tells us that in the thirteenth century, Robert the Bruce, discouraged and facing defeat in his war against the English, gained new resolve while watching a spider's persistent efforts to build a web in a difficult corner. When it finally succeeded, he decided to try one more time. He, too, succeeded.

The Benefits of Spiders

Most spiders are much more helpful than most of us give them credit for. In fact, they are among nature's most effective insecticides. According to Dr. Willis Gertsch, former curator of the American Museum of Natural History, spiders are probably far more important than birds in the number of insects they destroy. Lots of people are sentimental about birds, but who melts with tenderness over an eight-legged spinner? Yet these reclusive, nearsighted, extremely timid creatures are essential to farms and forests. They help to control flies, cotton pests, gypsy moths, and pea aphids, among many other destructive pests. They are especially important in cultivated cotton and corn fields, where crops are most vulnerable to ravenous insects. Quickly overrunning the fields when the pests emerge, spiders consume large numbers of larvae and adult insect pests.

Spiders are also most useful in cities. In 1925, crab spiders stopped a huge infestation of bedbugs in Athens, Greece. The same species is often found in New York City's food warehouses, preying on stored-food pests. Other spiders kill moths and grasshoppers. The wheel net spider, which destroys an average of 2,000 insects in its eighteen-month life, is used in Germany to protect forests.

Spiders carry no known diseases. Indeed, Italian peasants believe that cobwebs in the stable are directly related to cattle health. There is probably more than a little truth to this, because trapped in the skeins are stable flies, houseflies, gnats, mosquitoes, and other health menaces.

Most spiders pose no threat to humans. The fangs of the garden spider are so tiny that they cannot pierce our skin at all. In fact, only a handful of spider species in the United States are capable of biting people. If spider venom is needed for experimental purposes, the creature often has to be forced to bite, sometimes with electric shocks.

And spiders have some remarkable talents. Crab spiders can change their color to blend with flower petals, concealing themselves from their prey. The bolas spider is the original cowboy, throwing out a line weighted at the end with sticky liquid silk to rope in its prey. Like insects, spiders have the ability to regenerate an amputated leg.

Probably the most marvelous thing about spiders is also the most common—their webs. Whether as delicately elegant as the garden spider's or as messy as the black widow's, a spider's web is an engineering marvel, precisely suited to catching its builder's daily bread. Some build aerial webs; some build retreats on the ground; some fasten their webs under rocks or logs; and some, like the bolas spider, have just a single strand to catch their prey. Meticulously built, the garden spider's orb web takes about an hour to weave, a task the methodical creature does every morning, since the old one is spoiled by dead insects, dust, and fallen leaves. The black widow's untidy home always has a tunnel in which the timid creature can hide.

Spiders' silk—the substance from which they construct their webs—is as remarkable as the webs themselves. Once used in ancient China as sewing thread, the fibers range in thickness from twenty-millionths to one-millionth of an inch. The fibers' tensile strength is said to equal that of fused quartz, yet each fiber can be so fine that human technology cannot match it. The fibers make excellent markers for surveying and laboratory instruments. The cross hairs of the Norden bombsight used by the Allies in World War II were made from the silk of the black widow, as were sighting scopes for several models of military rifles.

Since the late 1980s, scientists have been investigating exciting new uses for spider silk, and have been trying to develop a useful synthetic. It may have potential for use in protective gear for soldiers and police. Researchers at the University of California, Santa Barbara, have recently discovered two significant qualities of synthetic spider silk: The molecules actually form strands on their own when mixed together, and the silk when stretched slowly returns to its original form. A private Canadian company is producing manmade spider silk using the mammary glands of goats, and German researchers have been producing the silk from, of all things, potatoes. Practical uses are probably still some years away but, like nylon, the twentieth century's miracle fiber, synthetic spider silk may be the wonder fiber of the twenty-first.

You may ask, "Why not just put a bunch of spiders together and let them do their thing without our interfering? We do that with silkworms, don't we?" The big difference is that silkworms aren't cannibals. Given the opportunity, spiders, who are carnivores, will kill and eat their fellows.

In addition to potential benefits of spider silk, researchers at the University of California, Riverside, have found that a protein in certain spider venoms may have a wide range of medical uses. It may have applicability in the manufacture of surgical implants, such as artificial heart valves and veins. Venoms are also being investigated as potential antitumor agents, anaesthetics, specialized natural pesticides, and anticonvulsants. In fact, it is thought that the barely tapped secrets of lowly spider venom may one day spur a host of high-tech medical breakthroughs.

The Hazards of Spiders

Although most spiders are harmless and generally helpful for human purposeś, there are two fairly widespread types whose bites are dangerous: the black widow and violin spiders. (See Figures 13.1 and 13.2.) Both are shy and nocturnal. Those most likely to suffer a severe reaction to their bites are very young children, older adults, and people in poor physical condition.

Table 13.1 summarizes these spiders' characteristics and their hiding places. Be careful around *any* place or thing that has not been disturbed for a long time. Picking up a plumber's helper once, I was startled by a sooty widow scurrying out from the plunger. (She was probably more scared than I was.)

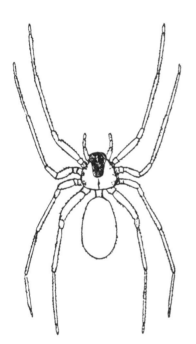

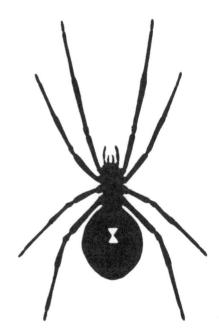

Figure 13.1
The Violin, or Brown Recluse, Spider
Source: Household and Stored-Food Insects of Public Health Importance and Their Control (Centers for Disease Control, Homestudy Course 3013-G, Manual 9, 1982.)

Figure 13.2
The Black Widow Spider
Source: Courtesy of James M. Stewart of the Centers for Disease Control, Centers for Infectious Diseases, Atlanta, Georgia.

	TABLE 13.1 CHARACTERISTICS OF THE BLACK WIDOW SPIDER AND THE VIOLIN SPIDER	
Characteristic	**Black Widow Spider**	**Violin Spider**
Color	Shiny black, sometimes gray.	Varies from light fawn to dark brown.
Size	Body $1/2$ inch long, globular abdomen.	Body $1/4$ to $1/2$ inch long, long legs.
Markings	Usually red hourglass on abdomen.	Possible "violin" on back; scroll toward rear.
Signs of Infestation	Webs; spider itself.	Long-legged, shriveled cast-off skins; webs; spider itself.
Web	Tangled, coarse, on or near ground in dark places; can trap a small mouse.	Above floor; white, cottony, with bluish support strands; usually with egg cases near floor; long, wide, multilayered bands.
Hiding Places	In barns, chicken coops, garages, cellars, outdoor privies, gas and water meters, angles of doors and windows, woodpiles, rubbish, old tires, grape and tomato vines; behind shutters.	In boxes, among papers, under rocks, in old clothes, in bedding lying on floor, on undersides of tables and chairs, in corners, behind pictures, at tops of window shades, under tree bark, in bird skeletons.

The Black Widow Spider

Despite its fearsome reputation, the black widow spider kills considerably fewer people than do snakes. Between 1716 and 1943, there were 1,300 cases of black widow bites in the United States, with 55 deaths recorded. Each year, by contrast, about 1,500 poisonous snake bites are reported in the country and, on the average, 75 of the victims die. In other words, in the United States, more people die from snake bites in one year than died from black widow bites in 227 years.

We can clean up their reputations a bit, but, with venom fifteen times more powerful than that of a rattler (but, fortunately, in comparably minute quantities), a black widow is nothing to take lightly. Control and precautions are essential. English zoos keep this species on exhibit, but when the international situation darkens, the colonies are destroyed so that no stray bomb can release them.

"The bite of the black widow spider need never be fatal," says entomologist Walter Ebeling, "if treated promptly by a physician." However, *anyone who has been bitten should receive antivenin at a medical facility as soon as possible.* Since the venom acts very quickly, cutting the wound open and sucking it does little good.

Try to kill the biting spider without squashing it beyond recognition, and give it to the doctor. Treatment for the wrong spider bite could be harmful.

Victims under sixteen or over sixty years old and those with high blood pressure should be hospitalized right away. Young children should be watched closely during their first ten hours of hospitalization. Without medical treatment, a child weighing less than thirty pounds has a 50-percent chance of dying from a black widow's bite. If you live in an area where these spiders abound, such as the southwestern United States, you should vacuum for spiders every two weeks in summer, and once a month in winter, and keep a close eye on toddlers who love to explore.

Symptoms of a black widow's bite are:

- Intense pain, first at the wound site, and then in the abdomen and legs
- Boardlike rigidity of the abdomen

The bite can also cause:

- Nausea and vomiting
- Faintness and dizziness
- Tremors
- Loss of muscle tone
- Shock
- Paralysis
- Speech disturbances
- Fever

In addition to the antidote, treatment may consist of mild sedation, muscle relaxants, and hot baths. Of course, the specific treatment regimen should be left to a knowledgeable medical doctor. *In no case should the patient be given any alcoholic drinks.*

Individual reactions to spider bites vary greatly. Untreated bites can lead to serious complications such as erysipelas (a skin infection), tetanus, and cerebral hemorrhage.

It may be some comfort to know that, "Despite its severe symptoms, spider bite poisoning is, in a majority of cases, a self-limiting condition, and generally clears up spontaneously within a few days," according to Dr. Emil Bogen.

The Violin Spider

Two kinds of violin spiders, the brown recluse and the *laeta,* are found in the United States. The brown recluse is fairly widespread in the Midwest; its bite, though serious, is rarely fatal. The more dangerous *laeta* first showed up in the United States in 1965, when a large infestation was discovered at Harvard University in Cambridge, Massachusetts. Five years later, violin spiders surfaced in southern California. They have also been seen in Ontario and British Columbia. The species are native to South America, where, up to 1968, they had bitten 400 times and caused at least 35 recorded deaths.

The California infestations were spread in cartons of merchandise. One clutch of them was in clothing that was slated for a church rummage sale.

Although the violin spider's venom can have very serious consequences, the animal is reluctant to use its fangs. It will attack, however, if caught in clothing one has just put on or if rolled on in a bed.

Today, doctors use corticosteroids to treat recluse spider bites, and with good results, but *it is essential that anyone bitten by this spider get to a physician as soon as possible*. Meanwhile, put ice on the wound. If possible, kill the spider without mutilating it, and take it along to the doctor so that it can be positively identified. Few physicians are trained to identify spiders, so if confronted with a spider-bite victim, they should be urged to have an expert confirm their identification. Research by entomologist Rick Vetter of the University of California, Riverside, has shown that violin spider bites are exceedingly rare in California and that most reports in that state are misdiagnosed bites that should be attributed to some other species. Misidentification could lead to ineffective treatment.

The violin spider poison causes a severe reaction at the wound site. This is different from the bite of the black widow, which causes a whole-body nerve reaction. The following are effects of violin spider bites:

- Gangrene develops around the puncture.

- The skin eventually sloughs off, leaving muscle and tendons exposed.

- Healing can take up to eight weeks.

- Heavy scarring can require plastic surgery.

- If the spider has injected the maximum amount of venom, the victim may suffer a systemic reaction, including blood and kidney problems.

SHUT THEM OUT

Because spiders can shut down their respiratory systems much more effectively than insects, they are relatively immune to most home pesticides. Prevention and mechanical controls are much more effective, and both are largely matters of good housekeeping and common sense.

Outdoors

❏ Keep your garden clean and neat.

❏ With a stick or broom, break down any untidy webs you find.

❏ Plant or trim your shrubs well away from the house to allow sun and wind to reach the building.

❑ Make sure the basement and any space under porches are dry. Keep the foundation well caulked.

❑ If your children have a tire swing, paint the inside of it white so that any spiders in it will be clearly visible. Look it over carefully before it is used, and, as your youngsters get older, alert them to this danger.

❑ Examine play equipment made with hollow tubing or pipes. Look inside for spiders or their webs.

❑ Before using an outdoor privy or portable toilet, take a stick in with you and run it under the seat to knock off any creatures underneath. (This is a favorite hideout for black widows.) Teach your children to do this. A bite on the genitalia is very serious, and men are the most frequent victims. If you are spending an extended period in a rustic area without indoor toilets, paint the underside of the privy seat with creosote to repel spiders.

❑ Handle garbage as recommended in Chapter 6.

❑ Dispose of pet droppings promptly so that spiders will look elsewhere for their prey.

❑ Use yellow bulbs for outdoor lighting. They attract fewer insects and, so, fewer spiders.

❑ Keep storage areas reasonably tidy.

❑ Always wear sturdy work gloves when cleaning garages, tool sheds, and garden houses, or even when just rummaging about in them.

❑ Carefully inspect jackets, sweaters, or coveralls hanging in the garage or tool shed before putting them on. These may be a great convenience for whoever does the yard work, but spiders often slip into clothing hanging on a wall.

❑ If you suspect that black widows or brown recluse spiders are on your property, put on sturdy closed shoes and look for them at night with a flashlight. Step on any you find. Or you can set out glue boards, such as those used to trap mice, to catch nocturnal spiders.

Indoors

The British have a saying, "There are three spiders in each room of every house no matter how houseproud the owner." Nevertheless, some of the same measures that keep out other unwanted wildlife will also suppress spiders in your house. These include the following:

❑ To keep out spiders' prey, maintain tight screening of windows, doors, and vents.

❏ Inspect all firewood, plants, and cardboard cartons before bringing them inside, just as you would for cockroaches and carpet beetles.

❏ Sweep behind washers and dryers regularly to oust potential troublemakers, since these appliances give spiders ideal nesting conditions, and are dark and moist enough to attract unwary prey.

❏ Move heavy furniture at reasonable intervals so that you can clean beneath it. Keep bedding off the floor.

❏ Periodically remove and clean curtains, pictures, and luggage.

❏ Consider rearranging your furniture. In Chile, where violin spiders are a serious problem, scientists recommend keeping a bed at least eight inches from the wall so that wall-crawling spiders can't get to it.

❏ Wearing heavy gloves and preferably using a small portable machine, vacuum up any webs and egg cases. Set the dust bag outside as soon as possible in a closed garbage container, placed in direct sun. Or put the bag into a larger plastic bag and set it in your freezer for forty-eight hours.

❏ Shake out any garments and shoes that have been unworn for a long time before putting them on, and look down into the shoes' toes before slipping into them. Be especially careful with clothing that has lain on the floor.

❏ Brush off, never swat, anything you feel crawling over your arm, the back of your neck, or your face when you are in bed at night. Swatting guarantees a bite, possibly venomous, while brushing catches the small explorer by surprise and removes it before it has a chance to do any harm.

❏ A prime rule in preventing spider bites is: Never thrust your hand or foot where your eye can't see. A corollary is: Never poke your bare hand up a chimney flue.

WASPS

On a recent hot summer's day, a retired couple decided to spend a few hours at a nearby mountain lake. While unpacking their lunch, they became the center of attention of an aggressive clan of wasps, which tasted the couple's food and invaded their picnic basket. The two quickly packed up and left the area. Bathers at the same lake also were harassed by the irascible buzzers.

Types of Wasps

The jury may still be out on wasps as friend or foe, but it's best to keep them at a healthy distance. To decide whether the wasp you're nervously eyeing is a real menace, you need to know something about the family as a whole. The more than 20,000 species can be roughly divided into two groups—solitary and social.

Solitary Wasps—Usually Harmless

Solitary wasps can nest underground, in a nail hole, or in a pithy plant stem. There, they lay their eggs and stock the burrow with paralyzed insects on which the larvae feed.

Solitary wasps are usually no threat to us. Although some are armed with frighteningly long stingers, they use them almost exclusively for paralyzing the insects they drag into their nests. They often specialize in one kind of prey; for example, the cicada killer, so large it's also called the king hornet, preys only on cicadas. Another species preys exclusively on grasshoppers; another, only on flies.

Social Wasps—Watch Out!

Social wasps, intent on protecting their nests, are much more apt to sting those humans they regard as intruders. Depending on the species, wasps' nests, which can measure up to six inches or more across, are built in leaf litter, in clumps of sphagnum moss, in trees or shrubs, under eaves, on the ceilings of garages, in outside water heater closets, in playground equipment, or underground. Aerial colonies tend to be much smaller than those underground.

So-called "paper wasps" build nests suspended by a central pedestal; you can usually see the developing larvae in open cells of their nest. The other kind of social wasp that tends to bother people is the yellow jacket. Depending on the species, yellow jackets have either aerial or subterranean nests. In contrast to paper wasps, yellow jacket nests are completely surrounded with paper. You can't see the cells with the developing brood unless you remove the outer paper.

In cooler climates, the colonies die when the temperature falls below freezing. The only survivors are fertilized queens, who hibernate until spring, sometimes in human homes. If you find a wasp at a window in spring, either let her out or swat her with a fly swatter, or you will soon have a nest inside.

Bees, ants, and termites—all nesting insects—swarm when they are ready to start new colonies. Wasps do not. New communities are founded in the spring by the emerging queen. The new colonies never reuse an old nest. Once empty, the nest is no danger to you. However, the insects may build a new one nearby. Incidentally, according to California Cooperative Extension, most wasps are actually beneficial. In one study, social wasps were found to collect more than ten million flies, caterpillars, and other insect pests in a summer along just three hundred feet of stream bed. Think what they could do over a wider area. Unless the nest is located where someone is apt to come upon it unexpectedly, like around a school, campground, or home, and startle the wasps, it should be left alone.

The Benefits of Wasps

Despite their unpleasant dispositions, wasps are helpful to us. They kill filth flies, black widow spiders, and many agricultural pests. Without one particular mini-wasp, the Calimyrna and Smyrna fig trees would not bear fruit, and one of California's major industries would fold up.

The galls, or swellings, that wasps make in some trees provide us with excellent inks and dyes. Aleppo oak ink, produced by a wasp gall, was once required by the United States Treasury, the Bank of England, and other government agencies for use on official documents.

The Hazards of Wasps

In the United States, wasps kill several dozen people every year. You may not class wasps with rattlesnakes but, before desensitization treatments were developed, wasps killed more people than rattlers did. And wasps did it a lot faster. Of those fatally stung, 80 percent died within a half-hour of the insect's attack, while only about 17 percent of rattlesnake victims succumbed in under six hours.

So aggressive are some species of wasps that they have even been recruited for war. The Vietcong used these "winged guerrillas" to stop the advance of South Vietnamese troops.

Most wasps, according to entomologist and wasp authority Howard Ensign Evans, have to be handled pretty roughly before they will sting; and, he adds, "to be stung by one is about as likely as being struck by a falling acorn." Only female wasps sting. Unlike the bee, which dies after stinging once, the wasp can needle a victim repeatedly. Scientists tell us that an alarmed wasp gives off a chemical attractant, or pheromone, that summons its coworkers and brings them swarming to attack the source of danger. Therefore, it is critical that you not frighten the one that lands on you in its search for food.

Wasps are more irritable on rainy days; sunny, warm weather keeps them in a mellower mood. They are antagonized by bright colors, and so are more apt to go after someone wearing vivid clothing. In the fall, with their colonies dispersed and no nests to defend, the insects are sluggish and not so quick to fly off the handle. But they can attack if disturbed. Be careful while clipping hedges and shrubs that might have hidden wasp nests. And be careful when picking fruit, as wasps like to build nests in fruit trees.

Treatment for Wasp Stings

Unless you are allergic to their venom, the effects of a wasp's sting can vary from a mild prick to severe local pain. The following treatments for ordinary wasp stings have all been used with success:

❏ Wash the area with soap and water (the site or stinger may be dirty).

❏ Apply an ice pack or cold compress of witch hazel to relieve swelling and pain.

❏ Apply an antihistamine cream, such as Benadryl, within twenty minutes of the stinging.

❏ Make a paste of baking soda and water, and apply it to the affected area.

❏ Make a paste of ¼ tablespoon of meat tenderizer and one or two teaspoonfuls of water, and apply it to the affected area. This helps to break down the venom, making it harmless.

❏ Dab on a tiny amount of household ammonia or a product containing ammonia that is formulated specifically for use on insect bites.

❏ If you are stung in the mouth, minimize the swelling by sucking on ice or taking slow sips of cold water.

❏ Bathe the area with a mild solution of vinegar or lemon juice. This will neutralize the venom. Apply calamine lotion to soothe the area.

Some people are extremely sensitive to bee and/or wasp venom. For them, stings can be life-threatening. Severe swelling in other parts of the body, hives, nausea, vomiting, cramps, difficulty in breathing, dizziness, hives, and unconsciousness are all possible complications. Stings on the head and neck and multiple stings are more dangerous. Kidney failure as the body tries to rid itself of the poisons is a real possibility.

Since 1959, treatment has been available that can desensitize most of those who are intensely sensitive. In addition to this treatment, doctors often prescribe an antidote kit to be used in case of a crisis. The kit consists of an injector to deliver a dose of adrenalin to block the allergic reaction quickly. If you are extremely allergic to insect bites and stings, keep a kit with you at all times (it is a good idea also to have one in the glove compartment of your car and in a secure place in your home.) Be aware, though, that an adrenalin kit is not a cure; it just buys time until you can get medical attention.

If you have a hypersensitivity to stings, let your primary care physician know so that he or she can both prescribe an adrenalin kit and never wastes valuable time diagnosing a nonexistent condition. Investigate the desirability of wearing a medical alert bracelet so that if you have a reaction that leaves you unable to speak, the medics will know what they they are dealing with. There have been cases in which hypersensitive people have died and the attending doctors called the cause heart failure.

People who take medication for high blood pressure and who have shown allergic reactions to stings before are particularly vulnerable to anaphylactic shock, as the venom pushes blood pressure dangerously low.

If you or your child has ever had a strong reaction to an insect bite, consult a competent allergist about desensitization. Schools may not have an antidote kit available, or their legal restrictions may not permit them to use it. A child struggling with a violent reaction to a bite may have to be taken by ambulance to a hospital miles away—a drive that could prove fatal. Only about half of 1 percent of people are hypersensitive to wasp stings, but don't let that fool you. That small number translates to 1 person out of 200, or 10 out of 2,000. So you can see that around a school with several hundred youngsters, there could be several who are extremely sensitive to insect bites without even knowing it.

Scientists at the Center for the Integration of Applied Science in Berkeley, California, recommend that those allergic to bee and wasp stings, in addition to having themselves immunized, follow these guidelines:

❏ Avoid wearing perfumes or perfumed substances like hairspray, suntan oil, and aftershave lotion.

❏ Wear white or light-colored clothing.

❏ Walk around, not through, masses of flowers or flowering shrubs.

❏ Avoid outdoor areas of strong food odors, such as barbecues and open garbage cans.

Other authorities strongly recommend these additional precautions:

❏ Don't mow grass, clip hedges, prune trees, or scythe tall grass.

❏ Don't go barefoot or wear sandals.

❏ Don't wear floppy clothing, and keep long hair tightly bound up to avoid entangling these notoriously touchy insects.

❏ Discourage allergic youngsters from eating popsicles or ice cream cones, and drinking soda outdoors during the summer.

❏ Be wary when near garbage cans and rotting fruit.

SHUT THEM OUT

To hold down the number of wasps in your vicinity, be sure to:

❏ Keep outside garbage cans tightly covered.

❏ Dispose of any overripe or rotting fruit in plastic bags set inside garbage cans.

❏ Fill up any abandoned rodent burrows on your property, as wasps often nest in these.

❏ On picnics, have a small can of tuna or pet food with you. Set it open some distance from your party to lure scavenging wasps away from your meal. If one comes near, stand still. Don't swat at it. If possible, put a handkerchief between you and the insect. *Should one land on you, remain quiet until it flies away.* If you frighten it, it will signal her sisters to come to her aid.

❏ Be especially cautious if you are having soda or fruit juice with your picnic. Wasps like it as much as you do, and getting a mouthful of wasp along with the drink can be disastrous.

WIPE THEM OUT

Any wasps' nest that is near areas of human activity, like entryways and children's play equipment, should be destroyed—whether the nest is underground or above ground.

Underground Nests

The safest way to destroy an underground wasps' nest is to have a certified pest control operator do the job. If you don't want to do that, you can try blowing silica gel dust or Drione (silica gel plus pyrethrins) down the nest entrance at night. This method is relatively safe and works very well, and is preferable to destroying the nest with insecticide. Because the chemical works slowly, you are much more likely to be stung when using an insecticide. Keep a shovel and a good-sized cloth handy. Wear protective clothing, like a beekeeper's outfit, including rubberized gloves and hats with bee veils. Slick-surfaced garments are safer than cotton or wool, which can be easily penetrated by the wasps' stingers. Tie your trouser legs and jacket cuffs over your boots and gloves. Plan to do the job after dark or very early in the morning, when the insects will be in the nest. Approach the nest very quietly so as not to arouse the highly sensitive guard wasps or other occupants. This technique is relatively safe and works very well. Be sure to have a helper shovel dirt over the opening to the nest, and leave the area immediately afterward.

Aerial Nests

The safest way to destroy an aerial paper nest is with a pressurized bomb specifically designed for wasps. Follow these guidelines:

❏ Do not use a general household insecticide. These work too slowly, so you run a high risk of being attacked.

❏ Be sure to follow label directions closely.

❑ Entomologist Don Reierson recommends that you get as close as possible, and then begin spraying directly at the nest. Work your way backwards to about ten feet, spraying continuously. When at a safe distance, stop—and leave the area.

❑ Alternatively, stand eight to ten feet away, and let the bomb shoot its stream of liquid into the nest. If you can get the stream right into the nest opening, so much the better, but just soaking the paper cover will also kill the colony.

Mud Daubers' Nests

One aerial nest that is easy to control is that of the mud dauber, a type of solitary wasp that may be found in many types of locations, including around homes and other domestic structures. These nests have only one occupant, an egg-laying female. Wait until she is either just leaving or just entering the nest, and kill her with a fly swatter. Then destroy the nest by scraping it away with a long-handled screwdriver or other tool. During the winter, you can safely remove the nest without insecticides, because only nonflying immature forms of the insect will be present.

REDUCING WASP POPULATIONS

Scientists at the United States Department of Agriculture recommend a "fish, wetting-agent, water" trap as an effective way to reduce the number of wasps in the neighborhood. As shown in Figure 13.3, a raw fish is suspended above a tub filled with water plus some detergent. The skin on the sides of the fish is cut open to give the insects ready access to the flesh. Yellow-jacket workers usually take pieces that are too heavy for them, fall into the water, and drown. If dogs or cats try to steal the bait, cover the trap with chicken wire or hardware cloth.

Commercial traps are also available that work well against most wasp species. Intensive trapping can reportedly reduce or suppress a wasp population in a limited outdoor area, something to keep in mind for a wedding reception or other open-air party. Be sure to ring the perimeter of the party area so that the wasps are drawn away from the center of human activity.

SCORPIONS

Recently a woman called a radio talk show host speaking about safe pest controls and asked, "How can I get rid of scorpions?"

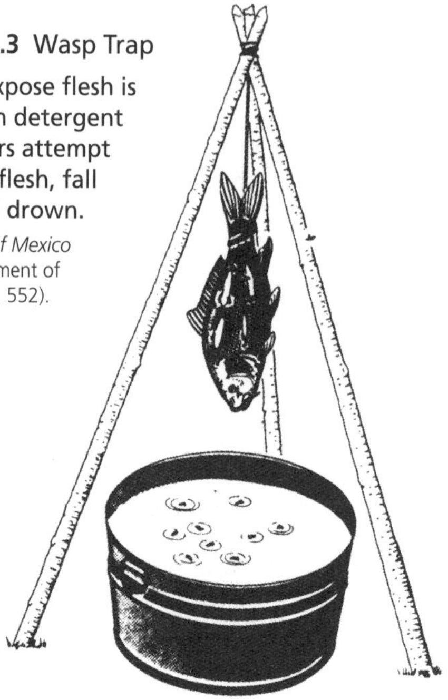

Figure 13.3 Wasp Trap

A raw fish with its sides cut to expose flesh is suspended over a pan of water to which detergent has been added. Yellow-jacket workers attempt to fly away with a large piece of flesh, fall into the water, and drown.

Source: The Yellow-Jackets of America North of Mexico (Washington, D.C.: United States Department of Agriculture Handbook p. 552).

"Where do you live?" the host asked.

"Scottsdale, Arizona," was her reply.

"Ma'am, you live in the desert and you have to expect to meet up with scorpions."

Somewhat indignant, the caller asserted, "I do *not* live in the desert!"

She was probably a refugee from an eastern suburb, and in Scottsdale had found a warmer, sunnier, largely artificial replica of her life back East. Enjoying all the conveniences of a metropolitan suburb, it had never occurred to her that the vastly different climate would have vastly different wildlife, some of which she had probably never seen before. There are as many as 650 different species of scorpions in the world, thirty of which are found in the United States.

Most scorpions that invade houses are no more dangerous than wasps (though, of course, if you are allergic to insect venom, that can be pretty dangerous). However, if you live in southern Arizona, southern New Mexico, or Mexico, the sculpturatus, or bark, scorpion is widespread, and its sting can be deadly. In one ten-month period, doctors treated more than 1,500 victims of scorpion stings; 233 of them had to be hospitalized. Sculpturatus is less than three inches long, but its sting can kill a small child. These scorpions are very apt to hide in children's sandboxes.

Figure 13.4 A Scorpion

Scorpions are difficult to eradicate because of their habit of hiding in unexpected nooks and crannies in the daytime. They will hide in shoes, blankets that brush the floor, clean folded bed sheets, piles of clothing. They come out at night, and have been known to drop onto a sleeper from the ceiling. They seek water, and are often found in kitchens and bathrooms.

Outdoors, they hide under rocks, trash piles, and under tree bark. When the weather turns wet (and at times it does rain in the desert), they are apt to show up in large numbers.

SHUT THEM OUT

The best ways to control scorpions are to close up any holes in your outside walls and cut tree branches away from the roof (scorpions can crawl along a branch to gain entry to an attic). You may need to hire a building contractor to do the thorough job needed.

Chemical pesticides are relatively useless against scorpions since these relatives of spiders do not nest, do not live in colonies, and are widely scattered. They can live for several days before dying of the poisoning and are still quite able to sting in the meantime. The best advice in combating scorpions is: *Be alert.* Because they tend to hide in the most unlikely places, you must inspect clothing before you put it on and also inspect bedding, whether it is neatly folded in a linen closet or on a bed. One entomologist, who should have known better, met up with a scorpion when he put on his undershorts in the dark.

If you live in scorpion country, the following are some steps you can take to prevent an assault:

❑ Remove piles of trash, rocks, lumber, and cardboard from near your house. Be sure to wear heavy gloves when doing this.

❑ If you live in the country, consider keeping chickens, ducks, or geese, all great scorpion hunters.

Scorpions in Medical Research

Medical researchers at the National Cancer Center at City of Hope in Duarte, California, have been investigating the tendency for scorpion venom to seek out and destroy brain cancer cells without harming healthy tissue. At this writing, eighteen patients with brain cancer have been chosen for clinical trials. The study is being carried out in collaboration with the University of Alabama, in Birmingham.

❏ If you have young children, *do not* give them a sandbox; it could become a favorite scorpion hangout.

❏ Because this relative of the spider cannot climb clean glass, set the legs of an infant's crib in clean, wide-mouthed glass jars.

WIPE THEM OUT

Start your own search and destroy mission. Lay damp—not wet—burlap or damp plywood sheets some distance from your house. Either of these will attract the thirsty arthropods. Shine a black light obtained from a scientific or biological supply company (not the type of black light used for parties) on the pests drawn to the trap. The scorpions will fluoresce a kind of chartreuse and you can step on them. It may take several weeks to trap the scorpions in your area, but your patience will be rewarded with your home's safety.

Although you may be grateful for the help that ants, spiders, wasps, and scorpions provide in suppressing the pests that can make you sick, you probably don't relish intimacy with any of them. By following the preventive steps recommended in this chapter, you will probably be able to keep these pest predators at a comfortable distance—where they will still be on on patrol.

14. Pesticides—
Only as a Last Resort

As you have read the preceding chapters, you have probably come to realize that you will rarely, if ever, need a chemical pesticide if you consistently follow the safe steps described. However, there are times when you need to get rid of unwanted wildlife in a hurry. With that in mind, this final chapter on using chemical pesticides safely is offered with a firm warning: *Be careful!*

THE USES OF PESTICIDES

Pesticides are not necessarily the evil that some people believe them to be. Sufficient quantities of basic foods like wheat, apples, tomatoes, carrots, and sweet corn could not be grown practically or economically without chemical pest controls. Pesticides also help to keep us healthy. Yellow fever, once a major killer in the southern United States, is almost unknown today, thanks primarily to public health authorities' spraying of mosquitoes. Millions of children around the world are growing up free of malaria and other insect-borne diseases because of the wide use of DDT in developing countries. Despite serious mistakes in the early disposal of DDT, it is estimated that in the first twenty-eight years of its existence, DDT prevented at least 20 million deaths and 200 million cases of disease.

The History of Pesticide Development

Modern pesticide use began in 1867, when Paris green, a combination of arsenic and copper, stopped the Colorado potato beetle's devastation of the United States potato crop. Around the same time, Bordeaux mixture, a blend of copper sulfate, lime, and water, was developed in France as a spray for grape pests. By the early twentieth century, fluorine compounds and botanical insecticides had come on the scene. The door to effective suppression of unwanted insects had been opened.

But the new substances also brought some problems. By the 1920s, the United States government was limiting the amount of lead and arsenic residues permitted on foodstuffs. And as early as 1912, long before the development of modern pesticides, scientists were noting insect resistance to chemical poisons. Among household pests that quickly show immunity, cockroaches, flies, and mosquitoes top the list. If only a few of the pests survive a poisonous dousing, those few pass their resistant genes to their offspring and a new generation of even tougher bugs is launched. Because pests develop quickly, a buildup of resistance to the point of control failure also develops quickly, sometimes within months.

To complicate matters, chemical pesticides usually also kill helpful insects like bees and, in some cases, cause overwhelming surges in what were once minor nuisances, such as spider mites. Although hundreds of pest species are now immune to some of our most powerful insect poisons, very few of the pests' natural enemies have developed any resistance to these chemicals. The chemicals are helping our foes and hurting our allies.

Modern Pesticide Development: A Slow, Expensive Process

No pesticide can be sold in the United States until the Department of Agriculture and the Environmental Protection Agency (EPA) have granted it registration, and their scrutiny takes six to eight years to complete. One major firm reports that of the 4,200 new compounds it tests each year, only two get even halfway through the screening process. Adding to the costs, in 1988 Congress ruled that all pesticides existing before 1984 had to be reregistered by 1997. Tests required to renew registration for just one pesticide can cost from $1 million to $12 million.

A company has to invest many millions of dollars before the item ever appears on market shelves. With up-front costs so high, pesticide manufacturers naturally push hard to sell their output. And householders, looking for easy pest control, are eager buyers—to the annual tune of billions of dollars.

Characteristics of Chemical Pesticides

There are 1,400 kinds of pesticides in the United States, and they are combined in 30,000 different ways. Some are termed "restricted-use" substances, and are meant to be applied only by licensed pest control operators against specific pests. Others are "general-use" products, considered by the government to be safe enough for use by the untrained public. *All pesticides, however, no matter what their category, are toxic.*

Chemical Categories of Pesticides

Chemically, there are five categories of pest poisons:

1. Chlorinated hydrocarbons (sometimes called organochlorines);

2. Organophosphates;

3. Carbamates;

4. Botanicals; and

5. Inorganic compounds.

Each group includes some that are deadly toxins and some that are fairly safe for humans *if used with care*. For example, heptachlor and chlordane, both dangerous, are chlorinated hydrocarbons. So is methoxychlor, generally considered one of the least risky pest controllers. All these chemicals were banned by the federal government several years ago because of their negative effects on humans and the environment Parathion, an organophosphate which is no longer used, can kill or maim for life; one of its relatives, malathion, is deemed safe enough to be sold to the general public and is used for ground and aerial control of the Mediterranean fruit fly. Strychnine is a botanical, and so are the pyrethrins, the safest of all pesticides. Deadly arsenic is an inorganic element, and the relatively mild boric acid is derived from borax, a naturally occurring mineral. So "natural" is not a synonym for "benign."

The chlorinated hydrocarbons, the organophosphates, and the carbamates all interfere with nerve functions. Some build up in mammalian tissues and eventually disrupt kidney and liver function. Pyrethrins, which are botanicals, may cause allergic reactions, and boric acid can bring on digestive disorders. None of these should be used nonchalantly.

Compounding the risk, the substance in which the insecticide is dispersed, usually a petroleum distillate, can be more toxic than the pesticide proper.

THE HAZARDS OF PESTICIDES

A Los Angeles woman who lived with her children and grandchildren had a fierce loathing for cockroaches. When a neighbor who worked as an uncertified applicator for a pest control company offered her some of his employer's leftover pesticide, she jumped at the chance to get rid of the roaches once and for all. Her friend may or may not have told her that the poison was full strength and should be applied only when greatly diluted. Ever the diligent housekeeper, the woman sprayed all the corners, baseboards, and closets of the home and then doused all the family's mattresses. As the babies, older children, and then the grownups began to feel sick, the adults decided to sleep on the lawn. Of course they pulled the mattresses outside, where Grandma, ever conscientious about protecting her family from contaminating bugs, sprayed the mattresses again. By the time the emergency-room doctor saw the family, all of them were seriously poisoned, the babies sickest of all.

Or you may have heard of the farmer who stored some leftover pesticide in a used cola bottle and set it in the refrigerator. His thirsty grandson came in, grabbed the bottle, and drained it . . . with tragic results. Stories not too different than these may be why you are reading this book.

According to the American Association of Poison Control Centers, in the year 2001 alone, 66,000 American children were victims of poisoning by common household pesticides. This number doesn't include another 28,000 poisoned by chlorine bleach.

In 1999, the Census Bureau estimated that 58 million households used insecticides, 14 million used fungicides, 40 million used herbicides, and 52 million used repellents.

And we constantly endanger ourselves and our families by our carelessness in using and storing these poisons. The U.S. Environmental Protection Agency (EPA) has found that 47 percent—nearly half—of all households with children under five years of age keep at least one pesticide in an unlocked cabinet less than four feet off the floor . . . within easy reach of preschoolers. And 75 percent of households without children store pesticides just as thoughtlessly. Incidentally, 13 percent of all childhood pesticide poisonings occur away from the child's home. You can never assume that other people, even relatives and friends, are as safety-minded as you are.

The "Cowboy Mentality" Problem

Entomologists believe that a poison pests carry back to their cohorts will wipe out a colony much more surely than a product that simply kills individual bugs on contact. We consumers, however, don't like to see an insect scamper away. Larry Newman, a vice president of S.C. Johnson, manufacturer of Raid insecticide, says, "Americans have a frontier mentality. The can of spray is their gun. They want the bug to die in front of them." So laboratories obligingly search for quick-kill formulas.

Scientists working for pesticide manufacturers and at universities claim that the bugs are winning our war with them. Their ability to withstand poisons and to develop immunity in their offspring is a powerful defense, one that stumps researchers over and over. Raid, perhaps the most widely distributed household pesticide in the United States, had its formula revised twenty-nine times between 1965 and 1984.This alone should give us pause when considering the use of pesticides.

The Carelessness Problem

Most of us are woefully ignorant when selecting and using a pesticide. When asked what they they are using, people usually answer "Raid" or "Black Flag," equating the chemical with its trade name.

The layperson's only source of information may be the retail salesperson in a nursery or hardware store, and many clerks are as uninformed as their customers of the real nature of what they are selling. Most of us either rely on ads or pick the most eye-catching container on the shelf.

The EPA has found that those who bother to read labels (fewer than half the buyers) are usually looking just for instructions on how to apply the chemical. Very few study the ingredients or antidotes or know much of anything about the product's components. Even fewer check with a county agent or college biology department to make sure that they have correctly identified the pest they are trying to control.

And when consumers get these potentially harmful products home, they use them as casually as they bought them. They turn a roach spray on flies, or an ant poison on fleas; they release a surface spray into the air, set off a patio fogger on a windy day, or hang a pesticide strip in an enclosed van.

Not surprisingly, according to the most recent statistics available, at least 82 percent of all pesticide poisonings happen at home, and children under age ten are the most frequent victims. When doctors see a family that is ill from exposure to organophosphates (a group of common antipest compounds), the babies are always the sickest, says Dr. Susan Tully, director of a major southern California emergency room. The smaller the person, the greater the effect. In addition, these substances tend to settle on floors and carpets, where crawling infants easily ingest them. Because they are small, their immune systems are immature, and their livers cannot detoxify poisons quickly, children react much more severely than adults to toxic substances.

There are only four ways to reduce the number of pesticide accidents:

1. Don't use chemical pesticides at all. Obviously, this is the surest way.

2. Learn as much as you can about any that you do use.

3. Apply those you use with great care.

4. Use the smallest amount that will do the job.

Pesticide Interactions with Other Substances

Scientists understand very little about the interaction of pesticides with one another or with drugs and food additives. Many tranquilizers, for example, are a form of chlorinated hydrocarbon and, like the larger group, act on our nerves. The relatively safe pyrethrins, derived from chrysanthemums, which may only stun an insect, are often combined with other poisons to make a much more toxic product.

USING AND STORING PESTICIDES SAFELY

Despite the complex chemistry and hazardous nature of pesticides, it is possible to choose them wisely and use them safely. The following sections of this chapter will outline how to do just that.

Identify the Pest Correctly

Your first step is to identify your target correctly. Do you really have silverfish, or are those small crawlers relatively harmless book lice? Do you have an invasion of earwigs, or is your home harboring some kind of cockroach? Unless you are positive that you know the species, catch a specimen in a small jar or box and check it against a book about insects at your public library. If you don't succeed there, take it to the biology department of a nearby college or high school. If these attempts lead nowhere, consult your county agricultural agent, a certified pest control operator, or a retail nursery.

Choose the Right Pesticide

Reading pesticide labels is the most important thing you can do in chemical pest control. Consider only those products that mention your target pest. It is illegal as well as unwise to choose a product designed for one type of pest for use against a totally different species.

Don't Look for Bargains

In 1997, New York State fined a well-known reputable pesticide manufacturer $950,000 for illegally selling a roach bait station with a powerful chemical that the EPA had banned as unsafe for use in the home. A market distributor had bought the dangerous product from the manufacturer and sold it to a Manhattan retailer specializing in obsolete and overstocked items.

Read Label Warnings

All poisonous substances in the United States must carry signal words alerting consumers to their danger. The degrees of potential harm indicated are as follows:

• *Danger—Poison* imprinted over a skull and crossbones. These are deadly. They kill quickly and violently. Ingesting anywhere from a tiny pinch to a teaspoonful is fatal to the average adult. They should rarely, if ever, be used.

• *Warning.* These are moderately toxic. Ingesting anywhere from a teaspoonful to a couple of tablespoonfuls will kill a normal person. The label may also say, "Harmful or fatal if swallowed." These should be used with extreme care.

- *Caution.* Somewhat less dangerous, but still toxic. From one ounce to one pint can be fatal. Use very carefully.

- *Keep Out of the Reach of Children.* The least poisonous. It could take as much as a quart to kills an average-sized adult.

Select Child-Resistant Packaging

Given a choice between an easily opened package and one that requires more effort, always choose the latter if you have children. Unlike drug manufacturers, pesticide companies are not legally required to offer their customers child-resistant packages for their products. Ideally, such closures should be compulsory on all hazardous substances, but the fact remains that they are not. (The only way to be absolutely certain that a container of pesticide is child-proof is to keep it where no child can possibly get to it.)

Asked if pesticide packages were safe, Dr. Bernard Portnoy, associate director of the Los Angeles County—University of Southern California Medical Center's pediatrics pavilion, answered with some vehemence, "They're not safe at all. They're always colorful. Think about how Raid is packaged and how it's advertised. It's something to *play* with. It sprays easily. It comes in beautifully colored cans."

Corinne Ray, formerly director of the Los Angeles Poison Information Center, adds, "Some of them you just poke a hole in the top and remove a piece of tape. It's impossible to reclose such a container securely. Once opened," she continues, "a pesticide package should have a very child-resistant lid on it."

What exactly is a child-resistant package? To test child resistance, the Consumer Product Safety Commission gives 200 children five minutes to open the product. Eighty-five percent of the youngsters must fail to open the package before being shown how; 80 percent must fail after being shown. Parents should lobby their representatives demanding these standards for pesticide packages. Write to manufacturers and let them know that you won't buy their product unless it is offered in a child-safe container. They will pay more attention to you than to a government agency. Also, buy only those products that a child can't open. Says Dr. Sue Tully, director of County–USC pediatric emergency room in Los Angeles, "If Black Flag makes several kinds of containers and the child-proof one sells the best, you'd better believe they're going to keep making that one."

Contact your nearest poison information center or hospital emergency room for stickers like those illustrated in Figure 14.1. Called "Mr. Yuk" by some, they give nonreaders vivid warning of the harmful nature of a container's contents.

Transport Pesticides Safely

Be aware of safety before putting a pesticide into your car. Never carry toxic sub-

Figure 14.1 "Mr. Yuk," either threatening or nauseated, warns non-readers of a container's dangerous contents. Poison center telephone number saves precious time in case of accidental ingestion.

Source: Insignias from Los Angeles County Medical Association Poison Center; Children's Hospital of Pittsburgh.

stances in the passenger compartment. Make sure the containers can't fall or be knocked over by a sudden stop. Protect paper bags from sharp objects, and both bags and boxes from moisture. And don't transport insecticides once they have been opened. They can spill and leave a lasting residue in your car.

Checklist for Using Toxic Substances Safely

❑ Read the entire label before buying the product.

❑ Choose the least toxic substance that will do your job.

❑ If you plan to use the poison indoors, make sure the label states that it's safe to do so.

❑ Never use indoors a pesticide that is designed for outdoor use.

❑ When you buy a pesticide for food plants, make sure the label says that it can be safely applied to the plant you are cultivating.

❑ If the label says to wear goggles, protective clothing, and a respirator while applying the product, do it, even if you have to buy the gear especially for the job. There are companies that specialize in safety clothing. The equipment will cost less than a hospital stay, time lost from work, or permanent damage to your health.

❑ Don't endanger other humans, pets, wildlife, livestock, beneficial insects, or ground water. Such hazards are often listed on the package.

❑ To minimize the chance that pesticide packages may deteriorate and develop leaks, buy only enough to do the job currently at hand, or only the amount you are sure you will use up in a few months.

❑ As a general rule, do not use any pesticide that is over five years old without contacting your local cooperative extension or Department of Agriculture for information on the current classification of older pesticides.

❑ Use the pesticide according to the directions on the label. Ignoring these instructions is a violation of federal law.

❑ Don't drink alcohol within twenty-four hours of using a pesticide.

❑ If you are pregnant or nursing, avoid any exposure to pesticides.

❑ Store and/or dispose of pesticides safely.

In addition to the general guidelines above, there are a number of precautions that apply specifically to certain situations.

Using Pesticides Outdoors

When applying pesticides indoors or out, safety must always be your first concern. The following precautions are suggested by the United States Department of Agriculture and/or the University of California Cooperative Extension:

❑ Work in an area with plenty of light and ventilation.

❑ Read the label's instructions thoroughly *before* opening the container.

❑ Make sure you wear *all* of the protective clothing mandated on the label, including shoes or boots. Always wear goggles or other protective eyewear.

❑ Use a sharp knife or other cutting tool to open paper sacks. *Never* tear them open.

❑ When opening a container of liquid, hold it well below eye level, with your face turned away from and to one side of the cap or lid.

❑ If you are using a concentrate, read the label to see what kind of rubber gloves you should wear while mixing the brew, because you will be handling it in its most dangerous form. There are different kinds of rubber gloves and the label will tell you the best ones for that product.

❑ Mix only the amount needed for the immediate job.

❑ Do not store mixed pesticides.

❑ Before filling a pesticide applicator, run clear water through it to check it for leaks, weakened hoses, loose connections, and pinholes in the tank. Repair any of these, or replace the equipment before mixing the pesticide.

❑ Use the product only for the pests and plants specified on the label.

❑ If it is a breezy day, postpone the application until the wind dies down. Always stand upwind of the applicator so that any unexpected puff of air blows the chemical droplets away from you.

❑ Do not apply pesticides near wells or cisterns, or on lawns when people or pets are around.

❑ After treating a lawn for soil pests, water the lawn to wash off the pesticide from the grass so there will be less hazard to children and pets.

❑ Keep children and animals away until the spray has dried.

❑ Avoid exposing bees to pesticides. Besides being beneficial, bees are extremely sensitive to some pesticides and may go berserk and sting anything in sight.

❑ Don't let pesticides drift onto your neighbors' property.

❑ Don't harvest pesticide-treated food before the interval specified on the label.

❑ Don't apply pesticides to bird nests or to blooming plants where pollinating insects may be working.

❑ *Don't smoke or eat while spraying.*

❑ Keep your hands away from your face while doing this work.

❑ Have safety and first aid equipment readily available.

Using Pesticides Indoors

❑ Assume that a prepared spray is flammable unless the label clearly says it is not.

❑ Turn off all gas pilots before starting to spray.

❑ Before spraying or dusting, clear the area of children and pets. Remove or cover all food. Take away pet food and water dishes.

❑ If using a surface spray, make sure it falls only on the surface you need to treat.

❑ Don't aim a spray into the room's air.

❑ Do not use a room fogger for spraying a surface. The droplets will remain in the air indefinitely.

❑ Never use any pesticide indoors that requires a gas mask or respirator for its application.

❑ Never spray entire walls, ceilings, or floors with a pesticide.

❑ If you are putting out baits, remember that these may be highly toxic. Set them where children and pets can't get at them.

Using Room Foggers

❑ Cover all food and cooking utensils before using a room fogger.

❑ Turn off all pilot lights.

❏ Close all windows and doors before spraying.

❏ Leave the room and close the door immediately after releasing the spray.

❏ Don't reenter the room until the time indicated on the label has elapsed.

Personal Safety and Toxic Exposure

❏ If you spill some pesticide while working, stop immediately. Assuming that none of the chemicals have splashed you, clean the material up right away. Sweep spilled dusts or powders into a container and dispose of it safely (see page 262). If you spill a liquid, soak it up with sand, sawdust., or clay-based cat litter and, after it has been absorbed, shovel it into a sturdy container and dispose of it properly. When washing spill areas, do not allow the wash water to flow into streams, ponds, or other bodies of water. If you are unsure of the best way to dispose of a spilled pesticide, call your county agent or local health department.

❏ Avoid accidental exposure. You can be contaminated by a pesticide through your mouth, your eyes, your respiratory system, and your broken or unbroken skin.

• *Oral exposure.* The most dangerous situation occurs when you get some of the material in your mouth. There, it can be absorbed through the tissues without ever reaching the rest of your digestive system. *Don't wait for symptoms to develop.* Get medical help as soon as possible. Take the pesticide container with you so that the chemical can be properly identified for correct treatment.

• *Eye exposure.* Get to a physician as soon as you can, but in the meantime start first aid right away. Hold the eyelid open and wash the eye for at least fifteen minutes with a gentle stream of clear, cool running water. Use no chemicals or drugs.

• *Respiratory exposure.* Go to a well-ventilated area at once, and don't go back to the work site until it is clear.

• *Skin exposure.* Wash the affected area thoroughly with soap and water. In both skin and lung exposures, if symptoms develop, get medical help. Again, take the container of pesticide with you. According to the National Academy of Sciences, poisoning in children through skin exposure is fairly common. In such cases, the youngster has usually been playing with an "empty" container of one of the more toxic pesticides. *From a health viewpoint, there is no such thing as an empty insecticide container.*

❏ Recognize signs of toxicity. *Acute pesticide poisoning* is a reaction that occurs soon after exposure to the substance. The symptoms may mimic those of other conditions, such as flu. Don't ignore them if there is any possibility of pesticide poison-ing. *Chronic poisoning* can result from repeated exposures over a long time. The symptoms can be similar to those of acute poisoning, but because of the time lag, they are much more difficult to detect. Exposure to some pesticides has been linked to cancer, birth defects, genetic mutations, and nerve disorders. With all the chem-ical substances we take into our homes (the EPA estimates that the average house-hold buys eighty different ones every year), every adult should be able to recognize common poisoning symptoms. Someone's life may depend on it. Here are the symptoms of pesticide poisoning:

- *Mild poisoning.* Symptoms may include fatigue, headache, dizziness, blurred vision, heavy sweating, nausea, vomiting, stomachache, diarrhea, and, possibly, heavy salivation.

- *Moderate poisoning.* The symptoms of mild poisoning worsen. You may not be able to walk, and may experience weakness, chest discomfort, muscle twitch-ing, and severe pupil constriction. Salivation may increase to the point of near-drowning.

- *Severe poisoning.* This may result in unconsciousness, difficulty in breathing, and convulsions, in addition to a worsening of the symptoms listed above. If untreated, death may ensue.

Other possible signs of exposure to toxins (not necessarily from pesticides), include:

- The presence of an open chemical container
- Chemical odor on the breath
- Abnormal behavior
- Hyperactivity or drowsiness
- Shallow or difficult breathing
- Burns on the mouth
- Loss of consciousness

Poisoning is much harder to spot in toddlers than it is in adults or older chil-dren. Anything, says pediatrician Sue Tully, can be a symptom. Be suspicious if your child *suddenly* becomes ill without having had a cold or "the general lousies" for a few days. Be alert, too, if a youngster *suddenly* starts throwing up. Unlike stomach flu, in poisoning, there will be more vomiting than diarrhea. If there are burns around the mouth, or if the child begins bumping into things or is unusually sleepy, *get in touch with a pediatrician immediately.*

The most dangerous time of day for childhood poisonings is from 5:00 to 9:00 p.m. Whoever prepares dinner is busy at the stove, children are hungry, and everyone is apt to be tired, so no one is watching the toddlers closely. Known as the "Children's Hour" in emergency rooms, this is the time to be especially watchful.

If you suspect poisoning, knowing what *not* to do may be more important than knowing what you should do. "Don't waste time with home remedies like milk or salt water," says Dr. Portnoy. "The most important thing to do is to get the victim to an emergency room as quickly as possible. Barring that, call a doctor."

Although there are poison information centers around the country, they may not want to talk to individual parents. A doctor is the one to contact them.

The most dangerous pesticides as far as children are concerned are those with the cholinesterase inhibitors, that is, those that affect the central nervous system. A child who has ingested one of these pesticides needs the fastest and most extensive treatment, according to Dr. Portnoy.

Do not try to induce vomiting unless a doctor tells you to. If a child has burns around the mouth or is drowsy, vomiting can add to the damage. However, every home with children should have syrup of ipecac on hand, to be given *only* if the doctor or poison information center advises it. Have enough of the medication for one ounce for each child in the home. Make sure the date on the bottle is current.

Antidotes listed on labels and first-aid charts should be used only in a pinch, if you are unable to contact any better source of advice. The problem with the antidotes listed on packages is that they may be completely out of date, especially if you have had the product around for a long time. In a random sample taken in 1982, the New York Poison Control Center found that 85 percent of toxic substances in 1,000 households had inadequate or totally wrong antidote information.

The common practice of inducing vomiting with salt water, frequently recommended, is particularly dangerous. Two tablespoons of salt can cause brain and lung damage as well as seizures in a small child. According to Dr. Barry Rumack, director of the Rocky Mountain Poison Center in Denver, Colorado, thirty to forty children died from salt poisoning during one ten-year period.

Cleaning Up

❏ Follow label directions.

❏ Rinse your equipment thoroughly at least three times with clear water.

❏ Drain the rinse water on unpaved ground in a sunny area, where it will not contaminate food plants or animals.

❏ Never pour rinse water down a plumbing drain or toilet.

❏ When eliminating pests chemically, whether indoors or outdoors, always wash any part of your skin that may have been exposed.

❏ If a chemical has fallen on your clothing, change clothes right away, and don't wear the garments again until they have been laundered.

❏ If your clothes have been heavily drenched by the pesticide, wear rubber gloves to pick them up, put them out in a plastic bag, and throw them out with the trash.

Storage

❏ Buy only enough pesticide for the short term. Store as little of these materials as possible. People may follow all the safety rules when using a pesticide and then stow it under the kitchen sink or with the medicines. Keep chemicals in a locked cabinet and always out of the view and reach of children. Check the containers from time to time to make sure they are not leaking.

❏ Application equipment should always be securely locked away or hidden, and must be labeled with the name of the chemical you use it for. Never use such equipment for more than one pesticide.

Proper Disposal

❏ According to the University of Massachusetts Cooperative Extension Service, you have three options in disposing of pesticides. You can send them to a land-fill, bury them diluted on your property, or bury them undiluted on your property. All of these options are undesirable. *None of these methods is acceptable for substances labeled "Danger—Poison," or for more than a few pounds or quarts of any toxic materials.*

Landfill Burial

❏ Send only a few containers to the landfill at a time. Do not puncture any that are pressurized. Wrap each container in several layers of newspaper so that it doesn't rupture in the trash truck and endanger the collectors. For large amounts, such as those accumulated during a community cleanup, hire a licensed pesticide hauler. Many communities now have an annual or semi-annual toxic trash collection day. If your community is among them, this is the time to get rid of your excess pesticides.

If you do not have trash disposal service, crush metal containers and break glass bottles. Then take the containers to the public dump, or bury them at least eighteen inches deep so that they will not contaminate any water source.

Never burn pesticide containers. You could release poisonous vapors.

Home Burial

If you must dispose of pesticide containers by means of home burial, you have two basic choices:

❑ *Dilute and bury.* This method is suitable for rural areas. Dilute leftover spray by adding about forty parts of water to one part of pesticide. For example, mix one-half quart leftover spray with five gallons of water. Pour this into a shallow trench in flat, well-drained woods or fallow fields with enough soil to absorb the liquid. Avoid sandy areas; bedrock, and steep slopes with runoff; as well as areas near water, gardens, or farms.

❑ *Bury undiluted.* Choose a site that will not be disturbed by excavation or construction *for at least five years.* Dig a hole eighteen to twenty inches deep. Put in two to three inches of charcoal followed by two to three inches of lime. Then pour in the pesticide. Do not bury unopened containers. (See Landfill Burial, above, for safe container disposal.) The chemical must have soil contact to decompose. Cover the pesticide with at least twelve inches of clean soil.

FUTURE PEST CONTROLS

Previous chapters have mentioned various discredited electronic antipest gadgets. Valueless as they are, these technologies may be the shadowy forerunners of more powerful future methods. Control techniques that show some promise include the following:

• Infrared light traps for specific pests.

• Radio frequency energy that kills stored-food pests without damaging the food.

• Devices that jam pests' communication signals.

• Organic pesticides that use the sun's ultraviolet rays to kill insect eggs and larvae. (These will not be a panacea for household pests, however, because many of those creatures flee from light.)

All of these controls are in the early testing stages, with many technical and economic problems to overcome.

As an increasingly aware public seeks safer, less poisonous substances to control household pests, industry has begun to offer pesticides that their marketers claim will do the job without endangering humans. Some of these products are being aggressively marketed in attractive packaging. It's too early to tell how effective they are, especially in the long run. If you are interested in learning of new, less toxic products and services coming onto the market, you can write to the Bio-Integral Resource Center, P.O. Box 7414, Berkeley, CA 94707. Along with their informative *Common Sense Pest Control Quarterly,* they publish an annual directory of least toxic products and services that will keep you up-to-date on this fast-changing field.

A FINAL WORD ABOUT PESTICIDE USE

With the flood of cleaning products and appliances available to American home-makers, one would think that our homes would be cleaner than in the past. Actually, they are dirtier. Women, the traditional housekeepers, have moved by the millions into the paid job market. With both marriage partners working, there is less time for housekeeping. Dust, clutter, forgotten spills, and carelessly handled food all favor insect life. To avoid the tedium of housecleaning, we turn to chemicals to do the job. For our own protection, let's handle these chemical "brooms" as infrequently as possible—and then only with the greatest caution.

References

Preface

Blondell, Jerome. *Human Pesticide Exposures Reported to Poison Control Centers in 2000.* Washington, DC: Health Statistics Division, Office of Pesticide Programs, US Environmental Protection Agency, December 2001.

General Accounting Office Finds Little Is Known About Pesticide Use in Schools and Health Risks to Children. Report requested by Senator Joseph Lieberman (Democrat of Connecticut), posted from http://www.senate.gov/member/ct/lieberman/general, January 4, 2000.

Koop, C. Everett. *American Council on Science and Health.* October 20, 1999.

1. Controls, Not Chemicals

Allman, William F. "Pesticides: An Unhealthy Dependence." *Science,* October 1985.

Ballantine, Bill. *Nobody Loves a Cockroach.* Boston: Little, Brown, 1968.

Bastien, Sheila, PhD. *Multiple Chemical Sensitivities (MCS): What It Is, What It Is Not, And How It Is Manifested.* Presentation at the Conference of the Association on Higher Education and Disability, 20 July 1995.

Bateman, Peter L.G. *Household Pests.* Poole, Dorset, England: Blandford Press, 1979.

Bellinger, Robert G. *Pest Resistance to Pesticides.* Clemson, SC: Clemson University Department of Entomology, March 1996.

Bottrell, Dale. *Integrated Pest Management.* Washington, DC: Council on Environmental Quality, 1979.

Carson, Rachel. *Silent Spring.* New York: Fawcett Crest, 1962.

Common Pantry Pests and Their Control. Berkeley, CA: Division of Agricultural Sciences, University of California. Leaflet 2711, 1969.

Dangerous Levels of Pesticides Found in Shellfish off Southern California Coast. Newscast. Los Angeles: KNXT (CBS affiliate), 20 September 1982.

Davidoff, Linda Lee. "Multiple Chemical Sensitivities (MCS)." *Amicus Journal,* Winter 1989.

Donaldson, Loraine. *Economic Development.* St. Paul, MN: West Publishing, 1983.

Ebeling, Walter. *Urban Entomology.* Berkeley, CA: Division of Agricultural Sciences, University of California, 1978 (rev.).

"Environmental and Economic Effects of Reducing Pesticide Use." *Common Sense Pest Control Quarterly,* Vol. 4 No. 3, Summer 1991.

Evolution, Science, and Society. Insect Pests: Resistance and Management. Stony Brook, NY: State University of New York at Stony Brook, undated.

Green, Albert, and Nancy L. Breisch. "Optimizing IPM for Buildings . . . " *Pest Control Technology,* July 2002.

Hartnack, Hugo. *202 Common Household Pests of North America.* Chicago: Hartnack, 1939.

"Hawaii's Urban Streams Severely Contaminated." *Earth Crash,* 3 January 2000.

Horowitz, Joy. "Common Pesticides: A Silent, Deadly Danger." *Los Angeles Times,* 29 April–1 May 1981.

Lacey, Mark S. "Meet the Beetles." *Pest Control Technology,* October 1996.

Levy, Phil. "Storing Grain: How to Keep the Bugs Out." *The Talking Food Newsletter,* April 1981.

Madon, Minoo. Biologist, California State Department of Health Services. Personal communication, 3 October 1983.

"School of Public Health Initiative Dedicated to Defeating Malaria." *Harvard University Gazette,* 20 January, 1999.

Sun, Marjorie. "Pests Prevail Despite Pesticides." *Science,* December 14, 1984.

Thomas, Hal, executive, Kamor Air Conditioning Inc., Pasadena, CA. Personal communication regarding dehumidifiers, 21 September 1983.

Tully, Sue, M.D., Director of Pediatrics Emergency Room, Los Angeles County–University of Southern California Medical Center. Personal communication, 29 September 1983.

Williams, Gene B. *The Homeowner's Pest Extermination Handbook.* New York: Arco, 1978.

2. Is Your Problem Really Insect Pests?

American Medical Association. *The Family Medical Guide.* New York: Random House, 1982.

Hartnack, Hugo. *202 Common Household Pests of North America.* Chicago: Hartnack, 1939.

Horowitz, Joy. "Common Pesticides: A Silent, Deadly Danger." *Los Angeles Times,* 29 April–1 May 1981.

New Columbia Encyclopedia. New York, London: Columbia University Press, 1975.

Olkowski, Helga, Bill Olkowski, Tom Javits, et al. *The Integral Urban House.* San Francisco: Sierra Club Books, 1979.

Scott, H.G., and J.M. Clinton. "An Investigation of 'Cable Mite' Dermatitis." *Annals of Allergy* 25 (1967): 409–414.

Shapiro, David. Director of Orange County [California] Poison Control Center. Interview, June 6, 1981.

"Short Takes." *Pest Control Technology*. November 1994.

Waldron, W.G. "The Role of the Entomologist in Delusory Parasitosis (Entomophobia)." *Entomological Society of America Bulletin* 8 (1962): 81–83.

Wilson, J.W. and H.E. Miller. "Delusion of Parasitosis (Acaraphobia.) *Archives of Dermatology* 54 (1946):39–56.

3. Outsmarting the Cagey Cockroach

Ballantine, Bill. *Nobody Loves a Cockroach*. Boston: Little, Brown, 1968.

Boraiko, Allen. "The Indomitable Cockroach." *National Geographic*, Vol. 159, No. 1, January 1981, pp. 130–142.

"Boric Acid for Cockroach Control." Berkeley, CA: Division of Agricultural Sciences, University of California. One-sheet answer #206, August 1967.

Brooke, James. "Sexual Lure for Roaches Synthesized." *The New York Times*, 26 September 1984.

"Cockroach Management." *The IPM Practitioner* 2(2), (February 1980).

"Cockroaches: Home and Landscape." *UC Pest Management Guidelines*. November 1999.

Controlling Household Pests. Washington, D.C.: United States Department of Agriculture. Home Garden Bulletin 96, 1971 (rev.).

Cooperative Extension, University of California, Davis. *Environment Toxicology Newsletter*, November 1999.

De Long, D.M. "Beer Cases and Soft Drink Cartons as Insect Distributors." *Pest Control* 30(7) (1962): 14–18.

Ebeling, Walter. *Urban Entomology*. Berkeley, CA: University of California, Division of Agricultural Sciences, 1978 (rev.).

"Fipronil—New Insecticide to Be Studied." *Quarterly Newsletter of the Center for Coastal Environmental Health and Biomolecular Research*, October 2000.

"From Killing Houseflies to Softening Our Ride: Virginia Tech Inventions and Creations Can Improve Our Lives." Press Release. Blacksburg, VA: Virginia Technical Institute, May 2001.

Hartnack, Hugo. *202 Common Household Pests of North America*. Chicago: Hartnack, 1939.

Hedges, Stoy. "Big Roaches, Big Jobs." *Pest Control Technology*, May 2002.

Maas, Rod. Inspector, Pasadena [California] Fire Department. Informal conversation, 2 April 1983.

Meyers, Charles. Biologist, California State Department of Health Services. Personal communication, 24 May 1983.

Olkowski, Helga, Bill Olkowski, and Tom Javits, et al. *The Integral Urban House.* San Francisco: Sierra Club Books, 1979.

Rachesky, Stanley. *Getting Bugs to Bug Off.* New York: Crown, 1978.

Ray, Corinne. Administrator of the Los Angeles County Medical Association Regional Poison Information Center. Personal communication, 1 June 1981.

Reierson, Don. Department of Entomology, University of California, Riverside. Personal communication, 9 January 1989.

Reierson, Don. Department of Entomology, University of California, Riverside. Personal communication regarding fipronil, 31 July 2002.

Slater, Arthur J. *Controlling Household Cockroaches.* Berkeley, CA: University of California Division of Agricultural Sciences Leaflet 21035, August 1979.

Williams, Gene B. *The Homeowner's Pest Extermination Handbook.* New York: Arco, 1978.

4. Critters in the Crackers, Pests in the Pantry

Ballantine, Bill. *Nobody Loves a Cockroach.* Boston: Little, Brown, 1968.

Bateman, Peter L.G. *Household Pests.* Poole, Dorset, England: Blandford Press, 1979.

Bush-Brown, James and Louise Bush-Brown. *America's Garden Book.* New York: Charles Scribner's Sons, 1947.

"Carpet Beetles: Home and Landscape." University of California Pest Management Guidelines, April 2001.

Common Pantry Pests and Their Control. Berkeley, CA: University of California Division of Agricultural Sciences Leaflet 2711, 1969.

Corrigan, Bobby. "Food Safety, IPM Guidelines." *Pest Control Technology,* July 2002.

Ebeling, Walter. *Urban Entomology.* Berkeley, CA: University of California Division of Agricultural Sciences, 1978 (rev.).

Flint, Mary, and Robert van den Bosch. *Introduction to Integrated Pest Management.* New York: Plenum, 1981.

Foster, Boyd. President, Arrowhead Mills, Hereford, TX. Personal communication, 31 October 1983.

Hartnack, Hugo. *202 Common Household Pests of North America.* Chicago: Hartnack, 1939.

Home and Garden Information Teletips. Los Angeles: University of California Cooperative Extension Service, undated.

Los Angeles County [California] Poison Control Center. Personal communication, 16 February 1990.

Olkowski, Helga, Bill Olkowski, and Tom Javits, et al. *The Integral Urban House.* San Francisco: Sierra Club Books, 1979.

Philbrick, Helen and John Philbrick. *The Bug Book*. Charlotte, VT: Garden Way, 1974.

Pratt, Harry D. *Mites of Public Health Importance and Their Control*. Atlanta, GA: Centers for Disease Control Homestudy Course 3013-G, Manual 8B, April 1976 (rev.).

Pratt, Harry D., Kent S. Littig, and Harold George Scott. *Household and Stored-Food Insects of Public Health Importance and Their Control*. Atlanta, GA: Centers for Disease Control Homestudy Course 3013-G, Manual 9, April 1976 (rev.).

Rachesky, Stanley. *Getting Pests to Bug Off*. New York: Crown, 1978.

Sutherland, Carol. "Spring Clean to Eliminate Pantry Pests." College of Agriculture and Home Economics, New Mexico State University, 1 March 1999.

Williams, Gene B. *The Homeowner's Pest Extermination Handbook*. New York: Arco, 1978.

5. Rats and Mice—There's No Pied Piper

Askham, Leonard R. "Ultrasonic and Subsonic Pest Control Devices." Washington State Univeristy Cooperative Extension Publications, January 1992.

Bateman, P.L.G. "The Rat, Exploiter of Man's World." *Pesticide Outlook*, 1991. Quoted in *Common Sense Pest Control Quarterly*, Winter 1996.

Biological Factors in Domestic Rodent Control. Atlanta, GA: Centers for Disease Control Homestudy Course 3013-G, Manual 10, April 1976 (rev.).

Bottrell, Dale. *Integrated Pest Management*. U.S. Government Printing Office, 1974.

Brooks, J.E. "A Review of Commensal Rodents and Their Control." *Critical Reviews in Environmental Control* 3(4): pp 405-453, 1973.

_____."Roof Rats in Residential Areas—The Ecology of Invasion." *California Vector Views* 13(9) (September 1966): 69–73.

Brooks, J.E., and A.M. Bowerman. "Anticoagulant Resistance in Rodents in the United States and Europe." *Journal of Environmental Health* 37(6): 537–542.

Canby, Thomas. "The Rat: Lapdog of the Devil." *National Geographic,* Vol. 152, No. 1, July 1977, pp. 60–87.

Daar, Sheila, Helga Olkowski, and William Olkowski. "Safe Ways to Solve Rat Problems In and Around Buildings." *Common Sense Pest Control Quarterly,* Winter 1986.

Dutson, Val J. "Use of the Himalayan Blackberry, *Rubus discolor,* by the Roof Rat, Rattus Rattus, in California." *California Vector Views* 20(8), August 1973, pp. 59–68.

Ebeling, Walter. *Urban Entomology*. Berkeley, CA: Division of Agricultural Sciences, University of California, 1978 (rev.).

Greaves, J.H., B.D. Rennison, and R. Redfern. "Warfarin Resistance in the Ship Rat in Liverpool." *International Pest Control* 15(2) (1973): 17.

Hartnack, Hugo. *202 Common Household Pests of North America*. Chicago: Hartnack, 1939.

The House Mouse: Its Biology and Control. Berkeley, CA: University of California Division of Agricultural Sciences Leaflet 2945, August 1981.

"IPM for Rats in Urban Areas." *Urban Ecosystem Management from The IPM Practitioner* 2(3) (March 1980).

Jamieson, Dean. "A History of Roof Rat Problems in Residential Areas of Santa Clara County, California." *California Vector Views* 12(6) (June 1965): 25–28.

Kavanau, J. Lee. "Behavior of Captive White-Footed Mice." *Science,* 31 March 1967.

Mackie, Richard A. "Control of the Roof Rat, Rattus Rattus, in the Sewers of San Diego." *California Vector Views* 11(2): (February 1964): 7–10.

Mallis, Arnold, ed. *Handbook of Pest Control,* 6th ed. Cleveland, OH: Franzak and Foster, 1982.

"Pets: Toy Fox Terrier." *Los Angeles Times,* 16 October 1983.

Philbrick, Helen and John Philbrick. *The Bug Book.* Charlotte, VT: Garden Way, 1974.

Plague: What You Should Know About It. Berkeley, CA: University of California Division of Agricultural Sciences Leaflet 21233, June 1981.

Pratt, Harry D., Bayard F. Bjornson, and Kent S. Littig. *Control of Domestic Rats and Mice.* Atlanta, GA: Centers for Disease Control Homestudy Course 3013-G, Manual 11, undated.

The Rat: Its Biology and Control. Berkeley, CA: Division of Agricultural Sciences, University of California. Fourth Edition. Leaflet 2896, 1981 (rev.).

Rats and Mice as Enemies of Mankind. London: British Museum (Natural History), Economic Series 15, undated.

Recht, Michael. Department of Biology, California State University, Dominguez Hills. Personal communication, 29 August 1985.

"Released Laboratory Rats as Community Pests in California." *California Vector Views* 18(10) (October 1971): 65–68.

"Rodent Control Prior to the Closing of Dumps." *California Vector Views* 18(12) (December 1971): 77–80

Subcommittee on Vertebrate Pests, National Research Council, National Academy of Sciences, eds. *Vertebrate Pests: Problems and Control.* Washington, DC: National Academy of Sciences, 1970.

Timm, R.M. "House Mouse." University of California Pest Management Guidelines, November 2000.

Virnig, Caron. "Getting the Bugs Out." *East West Journal,* May 1989.

6. The Fearsome Fly

Brown, Paul, Walter Wong, and Imre Jelenfy. "A Survey of the Fly Production from Household Refuse Containers in Salinas, California." *California Vector Views* 17(4) (April 1970): 19–22.

Chapman, John, and Dean H. Ecke. "Study of a Population of Chironomid Midges Parasitized by Mermithid Nematodes." *California Vector Views* 16(9) (September 1969): 83–84.

Ebeling, Walter. *Urban Entomology.* Berkeley, CA: University of California Division of Agricultural Sciences, 1978 (rev.).

Ecke, Dean H., and Donald Linsdale. "Control of Green Blow Flies by Improved Methods of Residential Refuse Storage and Collection." *California Vector Views* 14(4) (April 1967): 19–27.

Evans, Howard Ensign. *Life on a Little-Known Planet.* New York: E.P. Dutton, 1968.

Loomis, Edmond C. and Andrew S. Deal. *Control of Domestic Flies.* Berkeley, CA: University of California Division of Agricultural Sciences Leaflet 2504, May 1974 (rev.).

Madon, Minoo, and Charles Meyers. Biologists, Vector Biology and Control Branch, California Department of Health Services. Personal communication, 6 October 1983.

Magy, H.I., and R.J. Black. "An Evaluation of the Migration of Fly Larvae from Garbage Cans in Pasadena, California." *California Vector Views* 9 (1962): 55–59.

Olkowski, Helga, Bill Olkowski, and Tom Javits, et al. *The Integral Urban House.* San Francisco: Sierra Club Books, 1979.

Pratt, Harry D., Kent S. Littig, and Harold George Scott. *Flies of Public Health Importance and Their Control.* Atlanta, GA: Centers for Disease Control Homestudy Course 3013-G, Manual 5, 1975.

Ray, Corinne, Director, Los Angeles County Medical Association Regional Poison Information Center. Personal communication, 1 June 1981.

"Shoo Fly!" *Wellness Letter* [University of California, Berkeley] Vol. 5, No. 11, August 1989.

Shreck, C.E. Research entomologist, United States Department of Agriculture, Insect Repellent and Attractant Project, Gainesville, FL. Personal communication, 4 April 1984.

Spencer, T.S., R.J. Shimmin, and R.F. Schoeppner. "Field Tests of Repellents Against the Valley Black Gnat." *California Vector Views* 22(1) (January 1975): 5–7

Von Frisch, Karl. *Ten Little Housemates.* Trans. Margaret D. Senft. Oxford, England: Pergamon Press, 1960.

7. The Mosquito—A Deadly Nuisance

Abramson, Rudy. "Tires Stack Up as Hazard to the Nation." *Los Angeles Times*, 1 April 1990.

Amonkar, S.V., and A. Banerji. "Isolation and Characterization of the Larvicidal Principal of Garlic." *Science* 174 (1971): 1343–1344.

Ballantine, Bill. *Nobody Loves a Cockroach.* Boston: Little, Brown, 1968.

Bayer, Dr. Marc. Television interview. KNBC, Los Angeles, July 27, 1989.

"Biology and Control of *Aedes aegypti.*" Vector Topics No. 4. Atlanta, GA: Bureau of Tropical Diseases, U.S. Centers for Disease Control, September 1979.

Bottrell, Dale. *Integrated Pest Management.* Washington, DC: Council on Environmental Quality, 1979.

Brody, Jane E. "Mosquitoes: Why These Pests Find Us So Tasty and How to Stop Them." *Los Angeles Herald Examiner*, 17 July 1983.

Brown, A.W.A. "Attraction of Mosquitoes to Hosts." *Journal of the American Medical Association* 196 (1966): 249–252.

Carpenter, Stanley, and Paul Gieke. "Distribution and Ecology of Mountain Aedes Mosquitoes in California." *California Vector Views* 21(1–3) (January-March 1974): 1–8.

Challet, Gilbert. Director, Orange County Vector Control Agency. Personal communication, 10 October, 1983.

"Control of Western Equine Encephalitis." Vector Topics No. 3. Atlanta, GA: Bureau of Tropical Diseases, U.S. Centers for Disease Control, October 1978.

Ebeling, Walter. *Urban Entomology*. Berkeley, CA: Division of Agricultural Sciences, University of California, 1978 (rev.).

Evans, Howard Ensign. *Life on a Little-Known Planet*. New York: E.P. Dutton, 1968.

Garcia, Richard, Barbara Des Rochers, and William G. Voigt. "Evaluation of Electronic Mosquito Repellers under Laboratory and Field Conditions." *California Vector Views* 23(5,6) (May-June 1976): 21–23.

"Generically Engineered Vaccines Aim at Blocking Infectious Diseases in Millions." *The Wall Street Journal*, 25 October 1983.

Gray, Harold F. and Russel E. Fontaine. "A History of Malaria in California." Proceedings and Papers of the Twenty-Fifth Annual Conference of the California Mosquito Control Association. Turlock, CA, 1957.

Gutierrez, Michael C., Ernst P. Zboray, and Patricia A. Gillies. "Insecticide Susceptibility of Mosquitoes in California." *California Vector Views* 23(7,8) (July-August 1976): 2730.

Kramer, Marvin C. Executive Director, California Mosquito Control Association. Personal communication, 19 October 1983.

"Lyme Disease Alert." *University of California Wellness Letter*, Vol. 5, No. 9, June 1989.

Madon, Minoo, and Charles Meyers. Biologists, Vector Biology and Control Branch, California Department of Health Services. Personal communication, 6 October 1983.

"Malaria Spreads in New, Deadlier Forms." *Los Angeles Times*, 25 December 1983.

"Mosquitoes' Habits Bug Entomologists." *Los Angeles Times*, 10 April 1983.

Mosquitoes of Public Health Importance and Their Control. Atlanta, GA: Centers for Disease Control Homestudy Course 3013-G, Manual 6, 1969 (rev.).

Mulhern, Thomas. "An Approach to Comprehensive Mosquito Control." *California Vector Views* 19(9) (September 1972): 61–64.

Nasci, Roger S., Cedric W. Harris, and Cyresa K. Porter. "Failure of an Insect Electrocuting Device to Reduce Mosquito Biting." *Mosquito News* 43(2) (June 1983): 180–184.

New Columbia Encyclopedia. New York: Columbia University Press, 1975.

Prevention and Control of Vector Problems Associated With Water Resources. Ft. Collins, CO. Water Resources Branch, U.S. Centers for Disease Control, 1965.

Reeves, E.L., and C. Garcia. "Mucilaginous Seeds of the Cruciferae Family as Potential Biological Control Agents for Mosquito Larvae." *Mosquito News* 29 (1969): 601–607.

"Smithsonian Natural History Collections Critical in Fight Against Invasive Species." Smithsonian Research Reports No. 111, Winter 2003.

Soares, George, Kevin Hackett and William Olkowski. "IPM for Mosquitoes." *Urban Ecosystem Management from The IMP Practitioner* 2(6) (June 1980).

Stewart, James M. Atlanta, Centers for Disease Control, Center for Infectious Diseases. Personal communication, 13 October 1983.

Sunset Western Garden Book. Menlo Park, CA: Lane Magazine and Book Company, 1973 (rev.)

U.S. Centers for Disease Control. *Malaria Surveillance: Annual Summary 1987.* Atlanta, GA: U.S. Centers for Disease Control, 1988.

U.S. Centers for Disease Control. *Summary of Notifiable Diseases.* United States 1987 Morbidity and Mortality Weekly Report 36(57). Atlanta, GA: U.S. Centers for Disease Control, 1987.

Van den Bosch, Robert. *The Pesticide Conspiracy.* Garden City, NY: Doubleday, 1978.

Zimmerman, David. "The Mosquitoes Are Coming—And They Are Among Man's Most Lethal Foes." *Smithsonian* Vol. 14 No. 3, June 1983, pp. 28–38, June 1983. Follow-up letter in *Smithsonian* Vol. 14 No. 5, August 1983, p. 12.

8. The Mighty Flea, the Insidious Tick

Baker, Norman F. and Thomas B. Farver. "Failure of Brewer's Yeast as a Repellent to Fleas on Dogs." *Journal of the American Veterinary Medicine Association* 2 (51 July 1983): 212–214.

Ballantine, Bill. *Nobody Loves a Cockroach.* Boston: Little, Brown, 1968.

Bartlett, John. *Bartlett's Familiar Quotations,* 13th ed. Boston: Little, Brown, 1955.

Belfield, Wendell O, and Martin Zucker. *The Very Healthy Cat Book.* New York: McGraw-Hill, 1983.

Bottrell, Dale. *Integrated Pest Management.* Washington, D.C.: Council on Environmental Quality, 1979.

Buck, William B. "Clinical Toxicosis Induced by Insecticides in Dogs and Cats." *Veterinary Medicine/Small Animal Clinician,* August 1979, p. 1119

The California Flea Cycle (summer only). Moraga, CA: California Veterinary Medical Association leaflet. Undated.

Clarke, Anna P., DVM. "Pet Question and Answer." *Los Angeles Times,* 28 August, 1983.

Common Ticks Affecting Dogs. Berkeley, CA: Division of Agricultural Sciences, University of California Leaflet 2525 1978 (rev.).

Ebeling, Walter. *Urban Entomology.* Berkeley, CA: Division of Agricultural Sciences, University of California, 1978 (rev.).

Fleas, Fleas, Fleas. Miami, FL: Adams Veterinary Research Laboratories leaflet. Undated.

Fleas of Public Health Importance and Their Control. Atlanta, GA: Centers for Disease Control Homestudy Course 3013-G, Manual 7-A, 1973.

"Fleas Pursue Prey With Dogged Will." *Los Angeles Times*, 17 July 1983.

Frohbieter-Mueller, Jo. "Drown Those Fleas." *Mother Earth News*, July/August 1983.

Halliwell, Richard E.W. "Ineffectiveness of Thiamine (Vitamin B$_1$) as a Flea Repellent in Dogs." *Journal of the American Animal Hospital Association* 18 (May-June 1982): 423–426.

Hartnack, Hugo. *202 Common Household Pests of North America*. Chicago: Hartnack, 1939.

Health Facts. CBS TV, KNXT, Los Angeles, 27 May 1989.

Herb Products Company. North Hollywood, CA. Personal communication, 13 December 1983.

Jones, Terry. Chemist. Personal communication, 5 December 1983.

Keh, Benjamin. "The Brown Dog Tick." *California Vector Views* 11(5) (May 1964): 27–31.

Keh, Benjamin, and Allan M. Barnes. "Fleas as Household Pests in California." *California Vector Views* 8(11) (November 1961): 55–58.

Kunz, Jeffrey R.M., ed. *American Medical Association Family Medical Guide*. New York: Random House, 1982.

Lehane, Brendan. *The Compleat Flea*. New York: Viking, 1969.

"Lyme Disease: Don't Panic." *University of California Wellness Letter*, Vol. 4, No. 9, June 1988.

National Academy of Sciences. *Insect-Pest Management and Control*. Vol. 3, *Principles of Plants and Animal Pest Control*. Washington, D.C.: National Academy of Sciences, 1969.

New Columbia Encyclopedia. New York and London: Columbia University Press, 1975.

Olkowski, Helga, and William Olkowski. "New Products for Flea Management." *Common Sense Pest Control Quarterly*, Winter 1986.

Olkowski, Helga, Bill Olkowski, and Tom Javits, et al. *The Integral Urban House*. San Francisco: Sierra Club Books, 1979.

Olkowski, William, and Linda Laub. "IPM for Fleas." *Urban Ecosystem Management from The IPM Practitioner* 2(9) (September 1980).

Peavey, George, DVM. President of Southern California Veterinary Association. Personal communication, 29 August 1983.

Pratt, Harry D, and Kent S. Littig. *Ticks of Public Health Importance and Their Control*. Atlanta: Centers for Disease Control Homestudy Course 3013-G, Manual 8A, 1974.

Rachesky, Stanley. *Getting Pests to Bug Off*. New York: Crown, 1978.

Shreck, Cooperative Extension Research Entomologist, United States Department of Agriculture, Insect Repellent and Attractant Project. Personal communication, 4 March 1984.

Stewart, James. Entomologist, U.S. Centers for Disease Control. Personal communications, 1983–1984.

Von Frisch, Karl. *Ten Little Housemates.* Trans. Margaret D. Senft. Oxford, England: Pergamon Press, 1960.

Wellborn, Stanley. "It's Spring and Insects Are on the March." *U.S. News and World Report,* 28 May 1984.

White, Dee. Personal communicaiton, 6 December 1983.

Williams, Gene B. *The Homeowner's Pest Extermination Handbook.* New York: Arco, 1978.

9. Lice and Bedbugs—The Unmentionables

Ackerman, A.B. "Crabs—The Resurgence of Phthirus Pubis." *The New England Journal of Medicine* 278 (1968): 950–951.

Bottrell, Dale. *Integrated Pest Management.* Washington, DC: Council on Environmental Quality, 1979.

Carpet Beetles, Clothes Moths, Bedbugs, Fleas. East Lansing, MI: Michigan State University Cooperative Extension Service Pamphlet, 1969.

Ebeling, Walter. *Urban Entomology.* Berkeley, CA: University of California Division of Agricultural Sciences, 1978 (rev.).

Hartnack, Hugo. *202 Common Household Pests of North America.* Chicago: Hartnack, 1939.

Head Lice: An Alternative Management Approach. Berkeley, CA: John Muir Institute for Environmental Studies Center for the Integration of Applied Science Leaflet. Undated.

Howlet, F.M. "Notes on Head and Body Lice and Mosquitoes." *Parasitology* 10 (1917): 186–188.

Keh, Benjamin. "Answers to Some Questions Frequently Asked About Pediculosis." *California Vector Views* 26(5,6) (May-June 1979): 51–62.

Keh, Benjamin, and John Poorbaugh. "Understanding and Treating Infestations of Lice on Humans." *California Vector Views* 18(5) (May 1971): 23–30.

"Lice and Their Control." Berkeley, CA: University of California Division of Agricultural Sciences Leaflet 2528. Undated.

National Academy of Sciences. *Insect-Pest Management and Control.* Vol. 3, *Principles of Plants and Animal Pest Control.* Washington, D.C.: National Academy of Sciences, 1969.

New Columbia Encyclopedia. New York: Columbia University Press, 1975.

Nuttall, G.H.F. "The Biology of Pediculus Humanus." *Parasitology* 10 (1917): 18–85.

Olkowski, Helga, and William Olkowski. "Using IPM Principles for Managing Head Lice." *Urban Ecosystem Management from the IPM Practitioner* (1), (January 1980).

Olkowski, Helga, Bill Olkowski, and Tom Javits, et al. *The Integral Urban House.* San Francisco: Sierra Club Books, 1979.

Pratt, Harry D., and Kent S. Littig. *Lice of Public Health Importance and Their Control.* Atlanta, GA: Centers for Disease Control Homestudy Course 3013-G, Manual 7-B, 1973.

Progress: Newsletter of the National Pediculosis Association, April 1986, March 1987, and June 1987.

Rachesky, Stanley. *Getting Pests to Bug Off.* New York: Crown, 1978.

"'Tis the Season for Big Trouble from Tiny Lice." *Los Angeles Times,* 16 October 1983.

Wright, W.H. *The Bedbug: Its Habits and Life History and Methods of Control.* Washington, DC: Public Health Reports of the United States Public Health Service. Supplement No. 175, 1944.

10. Clothes Moths, Carpet Beetles, Silverfish, Firebrats, and Crickets

Carpet Beetles. London, England: British Museum (Natural History) Economic Leaflet No. 8, 1967.

Ebeling, Walter. *Urban Entomology.* Berkeley, CA: Division of Agricultural Sciences, University of California, 1978 (rev.).

Flint, Mary., and Robert Van den Bosch. *Introduction to Integrated Pest Management.* New York: Plenum, 1981.

Gradidge, J.M.G., A.D. Aitken, and P.E.S. Whalley. "Clothes Moths and Carpet Beetles." London, England: British Museum (Natural History) Economic Leaflet No. 14, 1967.

Hartnack, Hugo. *202 Common Household Pests of North America.* Chicago: Hartnack, 1939.

Insect Nuisances in Stores and Homes. London, England: British Ministry of Agriculture, Fisheries, and Food Leaflet GD52, May 1978.

Laudani, Hamilton. "PCO's Should Treat Cedar Chests and Closets for Moths." Reprinted by United States Department of Agriculture from *Pest Control*, October 1957.

Mallis, Arnold, ed. *Handbook of Pest Control*, 6th ed. Cleveland, OH: Franzak and Foster, 1982.

Moore, W.S., C.S. Koehler, and C.S. Davis. *Carpet Beetles and Clothes Moths.* Berkeley, CA: University of California Division of Agricultural Sciences Leaflet 2524, August 1979 (rev.).

Slater, Arthur and Georgia Kastanis. *Silverfish and Firebrats and How to Control Them.* Berkeley, CA: University of California Division of Agricultural Sciences Leaflet 21001, 1977.

Williams, Gene B. *The Homeowner's Pest Extermination Handbook.* New York: Arco, 1978.

11. Termites and Carpenter Ants—The Hidden Vandals

Ballantine, Bill. *Nobody Loves a Cockroach.* Boston: Little, Brown, 1968.

Bastin, Harold. *Insect Communities.* New York: Roy Publishers, 1957.

Bottrell, Dale. *Integrated Pest Management.* Washington, DC: Council on Environmental Quality, 1979.

Code of Federal Regulations, Title 40, Part 171.

Daar, Sheila. "Honolulu's Building Code OKs Termite Sand Barrier." *Common Sense Pest Control Quarterly,* Vol. 5 No. 3, Summer 1989.

Ebeling, Walter. "Electrogun Zaps Drywood Termites." *Pest Control Technology,* Vol. 13 No. 6, 1985.

____. "The Extermax System for Control of the Western Drywood Termite." Research paper published by ETEX, Las Vegas, Nevada. Undated.

____. *Urban Entomology.* Berkeley, CA: University of California Division of Agricultural Sciences, 1978 (rev.).

Ebeling, Walter., and Charles F. Forbes. "Sand Barriers for Subterranean Termite Control." *The IPM Practitioner,* Vol. 10 No. 5, May 1988.

Flint, Mary., and Robert Van den Bosch. *Introduction to Integrated Pest Management.* New York: Plenum, 1981.

Forbes, Charles F., and Walter Ebeling. "Update: Liquid Nitrogen Controls Drywood Termites." *The IPM Practitioner,* Vol. 8 No. 8, August 1986.

____. "Update: Use of Heat for Elimination of Structural Pests." *The IPM Practitioner,* Vol. 9 No. 9, August 1987.

Hall, Ron. "Turn up the Heat." *Pest Control,* February 1989.

Harris, W. Victor. *Termites, Their Recognition and Control.* New York: Wiley, 1961.

Hartnack, Hugo. *202 Common Household Pests of North America.* Chicago: Hartnack, 1939.

Heier, Albert. Public Information Officer, Environmental Protection Agency, Washington, DC. Personal communication, 27 January 1984.

Horowitz, Joy. "Common Pesticides: A Silent, Deadly Danger." *Los Angeles Times,* 29 April–1 May 1981.

"The Hot New Way to Banish Bugs." *Newsweek,* 12 February 1990.

"IPM for Termites." *Urban Ecosystem Management from The IPM Practitioner* 2(12), December 1980.

LaVoie, Robert E. President of Ace Termite and Pest Control Corporation, Los Angeles, CA. Personal communication, 2 February 1984.

Los Angeles County Poison Information Center. Personal communication, 21 February 1990.

Maeterlinck, Maurice. *The Life of the White Ant.* New York: Dodd, Mead, 1927.

Mahaney, J.H. "Termites on USNS *Sunnyvale,*" *Pest Control* 40(7) (1972): 17–18, 36, 38.

Mallis, Arnold, ed. *Handbook of Pest Control,* 6th ed. Cleveland, OH: Franzak and Foster, 1982.

National Academy of Sciences. *Insect-Pest Management and Control* Vol. 3, *Principles of Plants and Animal Pest Control.* Washington, D.C.: National Academy of Sciences, 1969.

Olkowski, Helga., and William Olkowski. "Carpenter Ants." *Common Sense Pest Control Quarterly,* Vol. 1 No. 2, Winter/Spring 1985.

Olkowski, Helga, Bill Olkowski, and Tom Javits, et al. *The Integral Urban House*. San Francisco: Sierra Club Books, 1979.

Prestwich, Glenn D. "Dwellers in the Dark: Termites." *National Geographic*, Vol. 153 No. 4, April 1978, pp. 532–546.

Rachesky, Stanley. *Getting Pests to Bug Off*. New York: Crown, 1978.

Rambo, George. Director of Technical Operations, National Pest Control Association. Personal communication, 6 February 1984.

Rust, Michael A., and Donald A. Reierson. *Termites and Other Wood-Infesting Insects*. Berkeley, CA: University of California Division of Agricultural Sciences Leaflet 2532, July 1981 (rev.).

Smith, Virgil K., H.R. Johnston, and Raymond H. Beal. *Subterranean Termites: Their Prevention and Control*. U.S. Department of Agriculture Bulletin No. 64, 1979.

Truman, Lee C., and William L. Butts. "Scientific Guide to Pest Control Operations." *Pest Control Magazine*, 1962.

Wilcox, W. Wayne, and David L. Wood. "So You've Just Had a Structural Pest Control Inspection." Berkeley, CA: University of California Division of Agricultural Sciences Leaflet 2999, July 1980.

World Book Encyclopedia, Vol. 16. Chicago: Field Enterprises, 1955.

12. Plant Pests—Part of the Landscape

California Plant Quarantine Manual. Amended September 2, 1988.

Carson, Rachel. *Silent Spring*. New York: Fawcett Crest, 1962.

Casual Home Invading Pests. East Lansing, MI: Michigan State University Cooperative Extension Service, 1969.

Doutt, R.S. *The Praying Mantis*. University of California Division of Agricultural Sciences Leaflet 21019, July 1980.

Dreistadt, Steve. "Gypsy Moth in California . . . Is Carbaryl the Answer?" *CBE Environmental Review*, May-June 1983.

Edwards, Dennis. Office of Pesticides Program, Environmental Protection Agency, Washington, D.C. Personal communication, 29 June 1989.

Huffaker, Carl B., ed. *New Technology of Pest Control*. New York: Wiley, 1980.

Long and Foster. *Home Trends*. Distributed by Herbert Hawkin Realty Co., June 1984.

McDonald, Elvin, Jacqueline Heritean, and Francesca Morris. *The Color Handbook of Houseplants*. New York: Hawthorn Books, 1975.

Mealybugs on Houseplants in the Home Landscape. Berkeley, CA: University of California Division of Agricultural Sciences Leaflet 21197, December 1980.

Mites in the Home Garden and Landscape, Berkeley, CA: University of California Division of Agricultural Sciences Leaflet 21048(B), July 1982 (rev.).

Moore, W.S. and C.S. Koehler. *Aphids in the Home Garden and Landscape*. Berkeley, CA: University of California Division of Agricultural Sciences Leaflet 21032, July 1979.

_____. *Earwigs and Their Control*. Berkeley, CA: University of California Division of Agricultural Sciences Leaflet 21010, August 1982.

_____. *Sowbugs and Pillbugs*. Berkeley, CA: University of California Division of Agricultural Sciences Leaflet 21015, July 1980.

Moore, W.S., and K.A. Hesketh. *Snails and Slugs in the Home Garden*. Berkeley, CA: University of California Division of Agricultural Sciences Leaflet 2530, August 1979 (rev.).

Moore, W.S., J.C. Profita, and C.S. Koehler. "Soaps for Home Landscape Insect Control." *California Agriculture* 33(6) (June 1979).

National Academy of Sciences. *Insect-Pest Management and Control*. Vol. 3, *Principles of Plants and Animal Pest Control*. Washington, D.C.: National Academy of Sciences, 1969.

"News From CBE's Minneapolis-St. Paul Office." *CBE Environmental Review*, May-June 1983.

"News You Can Use." *U.S. News and World Report*, 10 October 1983.

Philbrick, Helen, and John Philbrick. *The Bug Book*. Charlotte, VT: Garden Way, 1974.

Schwartz, P.H. *Control of Insects on Deciduous Fruits and Tree Nuts in the Home Orchard—Without Insecticides*. Washington, D.C.: United States Department of Agriculture Home and Garden Bulletin 211, May 1981 (rev.).

"Simple Tactics Against Your Slimy Foes." *Sunset Magazine*, April 1990.

Soares, G.G., Jr. "IPM for the Japanese Beetle." *The IPM Practitioner* 2(8) (August 1980).

West, Ron. "The Backyard Jungle: Spider Mites." *Mother Earth News*, July-August, 1983.

What to Do When the Bugs Come for Your Garden. Santa Barbara, CA: Community Environmental Council.

Whiteflies on Outdoor and Indoor Plants. Berkeley, CA: Division of Agricultural Sciences Division of Agricultural Sciences Leaflet 21267, November 1981.

Yepsen, Roger B., Jr., ed. *Organic Plant Protection*. Emmaus, PA: Rodale, 1976.

"Your Pest Control Program: One Safe Step at a Time." *Organic Garden and Farming* 22(4) (April 1975): 87–91.

13. Ants, Spiders, Wasps, and Scorpions—Useful but Unwelcome

Ants and Their Control. Berkeley, CA: Division of Agricultural Sciences, University of California. Leaflet 2526, reprinted September 1982.

Ants Indoors. London: British Ministry of Agriculture, Fisheries, and Food. Leaflet 366, 1969 (rev.).

Ballantine, Bill. *Nobody Loves a Cockroach*. Boston: Little, Brown, 1968.

Carson, Rachel. *Silent Spring*. New York: Fawcett Crest, 1962.

Control of Yellowjackets and Similar Wasps. Berkeley, CA: University of California Division of Agricultural Sciences Leaflet 2527. Undated.

Ebeling, Walter. *Urban Entomology*. Berkeley, CA: University of California Division of Agricultural Sciences, 1978 (rev.).

Evans, Howard Ensign. *Wasp Farm*. Garden City, NY: Natural History Press, 1963.

Evans, Howard Ensign, and Mary Jane West Eberhard. *The Wasps*. Ann Arbor, MI: University of Michigan Press, 1970.

Gertsch, Willis. *American Spiders*. New York: Van Nostrand Reinhold, 1979.

Gorham, J. Richard. *The Brown Recluse*. Washington, DC: United States Public Health Service Publication No. 2062, 1970.

_____. "The Geographic Distribution of the Brown Recluse Spider, *Loxosceles reclusa*, and Related Species in the United States." Cooperative Economic Insect Report (CEIR) 18 (1968): 171–175.

Hutchins, Ross E. *Insects*. Englewood Cliffs, NJ: Prentice-Hall, 1966.

Keh, Benjamin. "The Black Widow Spider." *California Vector Views* 3(1) (January 1956): 1, 3–4.

_____. "A Brief Review of Necrotic Arachnidism or North American Loxoscelism." *California Vector Views* 14(7) (July 1967): 48–50.

_____. "Loxosceles Spiders in California." *California Vector Views* 17(5) (May 1970): 29–34.

New Columbia Encyclopedia. New York: Columbia University Press, 1975.

Olkowski, Helga. "Aphids and Ways to Control Them on Plants Indoors." *Common Sense Pest Control Quarterly*, Vol. 3 No. 3, Summer 1987.

Olkowski, Helga, Bill Olkowski, and Tom Javits, et al. *The Integral Urban House*. San Francisco: Sierra Club Books, 1979.

Olkowski, William. "Ants in the Bay Area." John Muir Institute Memo, 1973.

_____. "The Spiders You Love to Hate." *Common Sense Pest Control Quarterly*, Vol. 3 No. 4, Fall 1987.

Philbrick, Helen, and John Philbrick. *The Bug Book*. Charlotte, VT: Garden Way, 1974.

Rachesky, Stanley. *Getting Pests to Bug Off*. New York: Crown, 1978.

Russell, Findlay E. "Bites of Spiders and Other Arthropods." In *Current Therapy*. Philadelphia: W.P. Saunders Co., 1969.

Stahnke, H.L. *The Treatment of Venomous Bites and Stings*. Tempe, AZ: Arizona State University, 1966.

Teale, Edwin Way. *The Strange Lives of Familiar Insects*. New York: Dodd, Mead, 1962.

U.S. Department of Agriculture. Agriculture Handbook No. 552. Washington, DC: U.S. Department of Agriculture, June 1981.

Vietmeyer, Noel. "Who Needs Spiders?" *Reader's Digest*, June 1989.

Von Frisch, Karl. *Ten Little Housemates*. Trans. Margaret D. Senft. Oxford, England: Pergamon Press, 1960.

Waldron, William G., Minoo B. Madon, and Terry Sudderth. "Observations on the

Occurrence and Ecology of *Loxosceles laeta* in Los Angeles County, California." *California Vector Views* 22(4) (April 1975): 29–35.

Wasps. London, England: British Ministry of Agriculture, Fisheries, and Food Leaflet GD53. 1978 (rev.).

Williams, Gene B. *The Homeowner's Pest Extermination Handbook.* New York: Arco, 1978.

World Book Encyclopedia. Chicago: Field Enterprises, 1955.

The Yellowjackets of America North of Mexico. Washington, DC: United States

14. Pesticides—Only as a Last Resort

Aikman, Lonnelle. "Herbs for All Seasons." *National Geographic,* Vol. 163 No. 3, March 1983, pp. 386–393.

Ballantine, Bill. *Nobody Loves a Cockroach.* Boston: Little, Brown, 1968.

Ballard, J.B., and R.F. Gold. "Ultrasonics—No Effect on Cockroach Behavior." *Pest Control,* June 1982.

Bennett, G.W., E.S. Runstrom, and J.A. Wieland. "Pesticide Use in Homes." *Bulletin of the Entomological Society of America,* Spring 1983.

Biological Factors in Domestic Rodent Control. Atlanta, GA: Center for Disease Control Homestudy Course 3013-G, Manual 10, 1976 (rev.).

Blondell, Jerome. "Pesticide-Related Injuries Treated in United States Hospital Emergency Rooms." *Calendar Year Report.* Washington, DC: U.S. Environmental Protection Agency Health Effects Branch, 1986.

Bottrell, Dale. *Integrated Pest Management.* Washington, DC: Council on Environmental Quality, 1979.

Children Act Fast—So Do Poisons. Washington, DC: United States Consumer Product Safety Commission Leaflet. Undated.

Ebeling, Walter. *Urban Entomology.* Berkeley, CA: University of California Division of Agricultural Sciences, 1978 (rev.).

"EPA Stops Sale of Several Electromagnetic Insect and Rodent Repellers." Press release. Washington, DC: U.S. Environmental Protection Agency, 1978 (rev.).

"Family's Dream Dashed by Bitter Irony." *Los Angeles Times,* 4 March 1984.

Greaves, J.H., and F.P. Rowe. "Responses of Confined Rodent Populations to an Ultrasound Generator." *Journal of Wildlife Management* (2) (April 1969): 409–417.

Howell, H.N., Jr., and T.A. Granovsky. *Report on Evaluation of Ultrasonic Pest Control Devices vs. American Cockroaches.* Unpublished. College Station, TX: Texas State University Department of Entomology. Undated.

Kostich, B.E. "Nature's Insecticide." *Life and Health Quarterly,* fourth quarter, 1980.

Kutz, Frederick W. "Evaluation of an Electronic Mosquito Repelling Device." *Mosquito News,* December 1974.

LaVoie, G.K., and J.F. Glahn. "Ultrasound as a Deterrent to Rattus Norvegicus." Denver, CO: Denver Wildlife Research Center, March 1976.

McCarron, Margaret M. "Acute Yellow Phosphorus Poisoning from Pesticide Pastes." *Clinical Toxicology* 18(6) (1981): 693–711.

Meehan, A.P. "Attempts to Influence the Feeding Behavior of Brown Rats Using Ultrasonic Noise Generators." *International Pest Control*, July-August 1976.

Nasci, R.S., C.W. Harris, and C.K. Porter. "Failure of an Insect Electrocuting Device to Reduce Mosquito Biting." *Mosquito News* 43(2) (June 1983): 180–184.

National Academy of Sciences. *Insect-Pest Management and Control*. Vol. 3, *Principles of Plants and Animal Pest Control*. Washington, D.C.: National Academy of Sciences, 1969.

"1987 Annual Report of the American Association of Poison Control Centers National Data Collection System." *American Journal of Emergency Medicine* 6(5) (September 1987).

"Pesticide Disposal for the Homeowner." *Pesticide Facts*. Boston: University of Massachusetts Cooperative Extension,undated.

Pesticide Toxicities. Berkeley, CA: University of California Division of Agricultural Sciences Leaflet 21062, 1979.

"Poison Antidotes Can Be Lethal Too." *Los Angeles Times*, 2 May 1983.

Poison Prevention Packaging. Washington, DC: U.S. Consumer Product Safety Commission Product Safety Fact Sheet No. 46, September 1980 (rev.)

Riker, Tom. *The Healthy Garden Book*. New York: Stein and Day, 1979.

Safe Use of Agricultural and Household Pesticides. Washington, DC: U.S. Department of Agriculture Handbook No. 321, 1967.

Shreck, C.E. Research entomologist, USDA Insect Repellent and Attractant Project. Personal communication, 4 March 1984.

Stead, Frank M. "Pesticides in Relation to Environmental Health." *California Vector Views* 11(1) (January 1964): 1–3.

Stimman, M.W., and J. Litewka. *Using Pesticides Safely in the Home and Yard*. Berkeley, CA: University of California Division of Agricultural Sciences Leaflet 21095, 1979.

"Sunlight That Kills." *Forbes*, 23 May 1983.

U.S. Environmental Protection Agency. *Highlights of the Findings of the National Household Pesticide Usage Study, 1976-1977*. Memo.

Zaslow, Jeffrey. "Scientists Seek Upper Hand in Insect Wars." *The Wall Street Journal*, 6 March 1984.

Index